The Sun to the Earth —and Beyond

A Decadal Research Strategy in Solar and Space Physics

Solar and Space Physics Survey Committee
Committee on Solar and Space Physics
Space Studies Board
Division on Engineering and Physical Sciences

NATIONAL RESEARCH COUNCIL
OF THE NATIONAL ACADEMIES

THE NATIONAL ACADEMIES PRESS
Washington, D.C.
www.nap.edu

THE NATIONAL ACADEMIES PRESS 500 Fifth Street, NW Washington, DC 20001

NOTICE: The project that is the subject of this report was approved by the Governing Board of the National Research Council, whose members are drawn from the councils of the National Academy of Sciences, the National Academy of Engineering, and the Institute of Medicine. The members of the committee responsible for the report were chosen for their special competences and with regard for appropriate balance.

Support for this project was provided by Contract NASW 96013 and NASW 01001 between the National Academy of Sciences and the National Aeronautics and Space Administration, National Oceanic and Atmospheric Administration Purchase Order No. 40-AA-NR-111308, National Science Foundation Grant No. ATM-0109283, Office of Naval Research Grant No. N00014-01-1-0753, and Air Force Office of Scientific Research Purchase Order No. FQ8671-0101168. Any opinions, findings, conclusions, or recommendations expressed in this material are those of the authors and do not necessarily reflect the views of the sponsors.

International Standard Book Number 0-309-08509-8 (Book)
International Standard Book Number 0-309-50800-2 (PDF)

Library of Congress Control Number 2003101592

Cover: The background photo is of the aurora borealis as viewed from the vicinity of Fairbanks, Alaska. The three figures in the inset show the magnetically structured plasma of the Sun's million-degree corona (left); the plasmasphere, a cloud of low-energy plasma that surrounds Earth and co-rotates with it (top right); and an artist's conception of Jupiter's inner magnetosphere, with the Io plasma torus and the magnetic flux tubes that couple the planet's upper atmosphere with the magnetosphere. Ground-based aurora photo courtesy of Jan Curtis; coronal image courtesy of the Stanford-Lockheed Institute for Space Research and NASA; plasmasphere image courtesy of the IMAGE EUV team and NASA; rendering of the jovian magnetosphere courtesy of J.R. Spencer (Lowell Observatory).

Copies of this report are available from the National Academies Press, 500 Fifth Street, N.W., Lockbox 285, Washington, D.C. 20055, (800) 624-6242 or (202) 334-3313 in the Washington metropolitan area. Internet, http://www.nap.edu

Copies of this report are available free of charge from:

Space Studies Board
National Research Council
500 Fifth Street, NW
Washington, DC 20001

Copyright 2003 by the National Academy of Sciences. All rights reserved.

Printed in the United States of America

THE NATIONAL ACADEMIES
Advisers to the Nation on Science, Engineering, and Medicine

The **National Academy of Sciences** is a private, nonprofit, self-perpetuating society of distinguished scholars engaged in scientific and engineering research, dedicated to the furtherance of science and technology and to their use for the general welfare. Upon the authority of the charter granted to it by the Congress in 1863, the Academy has a mandate that requires it to advise the federal government on scientific and technical matters. Dr. Bruce M. Alberts is president of the National Academy of Sciences.

The **National Academy of Engineering** was established in 1964, under the charter of the National Academy of Sciences, as a parallel organization of outstanding engineers. It is autonomous in its administration and in the selection of its members, sharing with the National Academy of Sciences the responsibility for advising the federal government. The National Academy of Engineering also sponsors engineering programs aimed at meeting national needs, encourages education and research, and recognizes the superior achievements of engineers. Dr. Wm. A. Wulf is president of the National Academy of Engineering.

The **Institute of Medicine** was established in 1970 by the National Academy of Sciences to secure the services of eminent members of appropriate professions in the examination of policy matters pertaining to the health of the public. The Institute acts under the responsibility given to the National Academy of Sciences by its congressional charter to be an adviser to the federal government and, upon its own initiative, to identify issues of medical care, research, and education. Dr. Harvey V. Fineberg is president of the Institute of Medicine.

The **National Research Council** was organized by the National Academy of Sciences in 1916 to associate the broad community of science and technology with the Academy's purposes of furthering knowledge and advising the federal government. Functioning in accordance with general policies determined by the Academy, the Council has become the principal operating agency of both the National Academy of Sciences and the National Academy of Engineering in providing services to the government, the public, and the scientific and engineering communities. The Council is administered jointly by both Academies and the Institute of Medicine. Dr. Bruce M. Alberts and Dr. Wm. A. Wulf are chair and vice chair, respectively, of the National Research Council.

www.national-academies.org

RECENT REPORTS OF THE SPACE STUDIES BOARD

Satellite Observations of the Earth's Environment: Accelerating the Transition of Research to Operations (2003)

Assessment of the Usefulness and Availability of NASA's Earth and Space Mission Data (2002)
Factors Affecting the Utilization of the International Space Station for Research in the Biological and Physical Sciences (prepublication) (2002)
Life in the Universe: An Assessment of U.S. and International Programs in Astrobiology (prepublication) (2002)
New Frontiers in the Solar System: An Integrated Exploration Strategy (prepublication) (2002)
Review of NASA's Earth Science Enterprise Applications Program Plan (2002)
"Review of the Redesigned Space Interferometry Mission (SIM)" (2002)
Safe on Mars: Precursor Measurements Necessary to Support Human Operations on the Martian Surface (2002)
Toward New Partnerships in Remote Sensing: Government, the Private Sector, and Earth Science Research (2002)
Using Remote Sensing in State and Local Government: Information for Management and Decision Making (prepublication) (2002)

Assessment of Mars Science and Mission Priorities (prepublication) (2001)
The Mission of Microgravity and Physical Sciences Research at NASA (2001)
The Quarantine and Certification of Martian Samples (prepublication) (2001)
Readiness Issues Related to Research in the Biological and Physical Sciences on the International Space Station (2001)
"Scientific Assessment of the Descoped Mission Concept for the Next Generation Space Telescope (NGST)" (2001)
Signs of Life: A Report Based on the April 2000 Workshop on Life Detection Techniques (prepublication) (2001)
Transforming Remote Sensing Data into Information and Applications (2001)
U.S. Astronomy and Astrophysics: Managing an Integrated Program (2001)

Copies of these reports are available free of charge from:
Space Studies Board
The National Academies
500 Fifth Street, NW, Washington, DC 20001
(202) 334-3477
ssb@nas.edu
www.nationalacademies.org/ssb/ssb.html

NOTE: Listed according to year of approval for release.

SOLAR AND SPACE PHYSICS SURVEY COMMITTEE

LOUIS J. LANZEROTTI, Lucent Technologies, *Chair*
ROGER L. ARNOLDY, University of New Hampshire
FRAN BAGENAL, University of Colorado at Boulder
DANIEL N. BAKER, University of Colorado at Boulder
JAMES L. BURCH, Southwest Research Institute
JOHN C. FOSTER, Massachusetts Institute of Technology
PHILIP R. GOODE, Big Bear Solar Observatory
RODERICK A. HEELIS, University of Texas, Dallas
MARGARET G. KIVELSON, University of California, Los Angeles
WILLIAM H. MATTHAEUS, University of Delaware
FRANK B. McDONALD, University of Maryland
EUGENE N. PARKER, University of Chicago, *Professor Emeritus*
GEORGE C. REID, University of Colorado at Boulder
ROBERT W. SCHUNK, Utah State University
ALAN M. TITLE, Lockheed Martin Advanced Technology Center

ARTHUR CHARO, Study Director
WILLIAM S. LEWIS,[1] Consultant
THERESA M. FISHER, Senior Program Assistant

---

[1] On temporary assignment from Southwest Research Institute.

PANEL ON THE SUN AND HELIOSPHERIC PHYSICS

JOHN T. GOSLING, Los Alamos National Laboratory, *Chair*
ALAN M. TITLE, Lockheed Martin Advanced Technology Center, *Vice Chair*
TIMOTHY S. BASTIAN, National Radio Astronomy Observatory
EDWARD W. CLIVER, Air Force Research Laboratory
JUDITH T. KARPEN, Naval Research Laboratory
JEFFREY R. KUHN, University of Hawaii
MARTIN A. LEE, University of New Hampshire
RICHARD A. MEWALDT, California Institute of Technology
VICTOR PIZZO, NOAA Space Environment Center
JURI TOOMRE, University of Colorado at Boulder
THOMAS H. ZURBUCHEN, University of Michigan

PANEL ON SOLAR WIND AND MAGNETOSPHERE INTERACTIONS

CHRISTOPHER T. RUSSELL, University of California, Los Angeles, *Chair*
JOACHIM BIRN, Los Alamos National Laboratory, *Vice Chair*
BRIAN J. ANDERSON, Johns Hopkins University
JAMES L. BURCH, Southwest Research Institute
JOSEPH F. FENNELL, Aerospace Corporation
STEPHEN A. FUSELIER, Lockheed Martin Advanced Technology Center
MICHAEL HESSE, NASA Goddard Space Flight Center
WILLIAM S. KURTH, University of Iowa
JANET G. LUHMANN, University of California, Berkeley
MARK MOLDWIN, University of California, Los Angeles
HARLAN E. SPENCE, Boston University
MICHELLE F. THOMSEN, Los Alamos National Laboratory

PANEL ON ATMOSPHERE-IONOSPHERE-MAGNETOSPHERE INTERACTIONS

MICHAEL C. KELLEY, Cornell University, *Chair*
MARY K. HUDSON, Dartmouth College, *Vice Chair*
DANIEL N. BAKER, University of Colorado at Boulder
THOMAS E. CRAVENS, University of Kansas
TIMOTHY J. FULLER-ROWELL, University of Colorado at Boulder
MAURA E. HAGAN, National Center for Atmospheric Research
UMRAN S. INAN, Stanford University
TIMOTHY L. KILLEEN, National Center for Atmospheric Research
CRAIG KLETZING, University of Iowa

JANET U. KOZYRA, University of Michigan
ROBERT LYSAK, University of Minnesota
GEORGE C. REID, University of Colorado at Boulder
HOWARD J. SINGER, NOAA Space Environment Center
ROGER W. SMITH, University of Alaska

PANEL ON THEORY, MODELING, AND DATA EXPLORATION

GARY P. ZANK, University of California, Riverside, *Chair*
DAVID G. SIBECK,[1] NASA Goddard Space Flight Center, *Vice Chair*
SPIRO K. ANTIOCHOS, Naval Research Laboratory
RICHARD S. BOGART, Stanford University
JAMES F. DRAKE, JR., University of Maryland
ROBERT E. ERGUN, University of Colorado at Boulder
JACK R. JOKIPII, University of Arizona
JON A. LINKER, Science Applications International Corporation
WILLIAM LOTKO, Dartmouth College
JOACHIM RAEDER, University of California, Los Angeles
ROBERT W. SCHUNK, Utah State University

PANEL ON EDUCATION AND SOCIETY

RAMON E. LOPEZ, University of Texas, El Paso, *Chair*
MARK ENGEBRETSON, Augsburg College, *Vice Chair*
FRAN BAGENAL, University of Colorado
CRAIG DEFOREST, Southwest Research Institute
PRISCILLA FRISCH, University of Chicago
DALE E. GARY, New Jersey Institute of Technology
MAUREEN HARRIGAN, Agilent Technologies
ROBERTA M. JOHNSON, National Center for Atmospheric Research
STEPHEN P. MARAN, NASA Goddard Space Flight Center
TERRANCE ONSAGER, NOAA Space Environment Center

---

[1] Johns Hopkins University Applied Physics Laboratory until summer 2002.

COMMITTEE ON SOLAR AND SPACE PHYSICS

JAMES L. BURCH, Southwest Research Institute, *Chair*
JAMES F. DRAKE, University of Maryland
STEPHEN A. FUSELIER, Lockheed Martin Advanced Technology Center
MARY K. HUDSON, Dartmouth College
MARGARET G. KIVELSON, University of California, Los Angeles
CRAIG KLETZING, University of Iowa
FRANK B. McDONALD, University of Maryland
EUGENE N. PARKER, University of Chicago, Professor Emeritus
ROBERT W. SCHUNK, Utah State University
GARY P. ZANK, University of California, Riverside

ARTHUR CHARO, Study Director
THERESA M. FISHER, Senior Program Assistant

NOTE: Members listed are those who served during the survey study period in 2001-2002.

SPACE STUDIES BOARD

JOHN H. McELROY, University of Texas at Arlington (retired), *Chair*
ROGER P. ANGEL, University of Arizona
JAMES P. BAGIAN, Veterans Health Administration's National Center for
    Patient Safety
ANA P. BARROS, Harvard University
RETA F. BEEBE, New Mexico State University
ROGER D. BLANDFORD, California Institute of Technology
JAMES L. BURCH, Southwest Research Institute
RADFORD BYERLY, JR., University of Colorado at Boulder
ROBERT E. CLELAND, University of Washington
HOWARD M. EINSPAHR, Bristol-Myers Squibb Pharmaceutical Research
    Institute
STEVEN H. FLAJSER, Loral Space and Communications Ltd.
MICHAEL FREILICH, Oregon State University
DON P. GIDDENS, Georgia Institute of Technology/Emory University
RALPH H. JACOBSON, The Charles Stark Draper Laboratory (retired)
MARGARET G. KIVELSON, University of California, Los Angeles
CONWAY LEOVY, University of Washington
BRUCE D. MARCUS, TRW, Inc. (retired)
HARRY Y. McSWEEN, JR., University of Tennessee
GEORGE A. PAULIKAS, The Aerospace Corporation (retired)
ANNA-LOUISE REYSENBACH, Portland State University
ROALD S. SAGDEEV, University of Maryland
CAROLUS J. SCHRIJVER, Lockheed Martin
ROBERT J. SERAFIN, National Center for Atmospheric Research
MITCHELL SOGIN, Marine Biological Laboratory
C. MEGAN URRY, Yale University
PETER VOORHEES, Northwestern University
J. CRAIG WHEELER, University of Texas at Austin

JOSEPH K. ALEXANDER, Director

Preface

The Sun to the Earth—and Beyond: A Decadal Research Strategy in Solar and Space Physics is the product of an 18-month effort that began in December 2000, when the National Research Council (NRC) approved a study to assess the current status and future directions of U.S. ground- and space-based programs in solar and space physics research. The NRC's Space Studies Board and its Committee on Solar and Space Physics organized the study, which was carried out by five ad hoc study panels and the 15-member Solar and Space Physics Survey Committee, chaired by Louis J. Lanzerotti, Lucent Technologies. The work of the panels and the committee was supported by the National Aeronautics and Space Administration (NASA), the National Science Foundation (NSF), the National Oceanic and Atmospheric Administration (NOAA), the Office of Naval Research (ONR), and the Air Force Office of Scientific Research (AFOSR).

The Sun to the Earth—and Beyond is the report of the Solar and Space Physics Survey Committee. It draws on the findings and recommendations of the five study panels, as well as on the committee's own deliberations and on previous relevant NRC reports. The report identifies broad scientific challenges that define the focus and thrust of solar and space physics research for the decade 2003 through 2013, and it presents a prioritized set of missions, facilities, and programs designed to address those challenges.

In preparing this report, the committee has considered the technologies needed to support the research program that it recommends as well as the policy and programmatic issues that influence the conduct of solar and space physics research. The committee has also paid particular attention to the applied aspects of solar and space physics—to the important role that these fields play in a society whose increasing dependence on space-based technologies renders it ever more vulnerable to "space weather." The report discusses each of these important topics—technology needs, applications, and policy—in some detail. *The Sun to the Earth—and Beyond* also discusses the role of solar and space physics research in education and examines the productive cross-fertilization that has occurred between solar and space physics and related fields, in particular astrophysics and laboratory plasma physics.

Each of the five study panels was charged with surveying its assigned subject area and with preparing a report on its findings. The first three panels focused on the important scientific goals within their respective disciplines and on the missions, facilities, programs, technologies, and policies needed to achieve them. In contrast, the Panel on Theory, Modeling, and Data Exploration addressed basic issues that transcend disciplinary boundaries and that are relevant to all of the subdisciplines of solar and space physics. The Panel on Education and Society examined a variety of issues related to both formal and informal education, including the incorporation of solar and space physics content in science instruction at all levels, the training of solar and space physicists at colleges and universities, and public outreach. The reports of the panels are published in a separate volume titled *The Sun to the Earth—and Beyond: Panel Reports* (2003, in press).

In addition to the input from the five study panels, the committee also received information at a 2-day workshop convened in August 2001 to examine in detail issues relating to the transition from research models to operational models. Participants in the workshop included members of the committee and representatives from the Air Force, the Navy, NOAA, NSF, NASA, the U.S. Space Command, academia, and the private sector.

The committee undertook its work intending to provide a community assessment of the present state and future directions of solar and space physics research. To this end, the committee and the panels engaged in a number of efforts to ensure the broad involvement of all segments of the solar and space physics communities. These efforts included town-meeting-like events held at the May 2001 joint meeting of the American Geophysical Union (AGU) and the American Astronomical Society's (AAS's) Solar Physics Division[1] and at spring and summer 2001 workshops of the following programs: International Solar-Terrestrial Physics (ISTP), Solar, Heliospheric, and Interplanetary Environment (SHINE), Coupling, Energetics, and Dynamics of Atmospheric Regions (CEDAR), and Geospace Environment Modeling (GEM). Each of these outreach events was well attended

---

[1]The AGU and the Solar Physics Division of the AAS are the two principal scientific organizations representing the solar and space physics community.

and provided the committee and panels with valuable guidance, suggestions, and insights into the concerns of the solar and space physics community. Additional community input came from presentations on science themes, missions, and programs at panel meetings, from direct communication with individual panel and committee members by phone and e-mail, and through Web sites and Web-based bulletin boards established by two of the panels. Reports in the electronic newsletters of the AGU's Space Physics and Aeronomy section and of the AAS's Solar Physics Division kept those communities informed of the progress of the study and encouraged their continued involvement in the study process.

Each of the study panels met at least twice during the spring and summer of 2001. The Panel on the Sun and Heliospheric Physics and the Panel on Education and Society met three times. The committee met five times, three times in 2001 and twice in 2002. The panel chairs and vice chairs participated in two of those meetings, during which they presented their panels' recommendations and received comments and suggestions from the committee. The final set of scientific and mission, facility, and program priorities and other recommendations was established by consensus at the committee's last meeting, in May 2002.

The committee's final set of priorities and recommendations does not include all of the recommendations made by the study panels, although it is consistent with them.[2] Each panel worked diligently to identify the compelling scientific questions in its subject area and to set program priorities to address these questions. All of the recommendations offered by the panels merit support; however, the committee took as its charge the provision of a strategy for a strong, balanced national program in solar and space physics for the next decade that could be carried out within what is currently thought to be a realistic resource envelope. Difficult choices were inevitable, but the recommendations presented in this report reflect the committee's best judgment, informed by the work of the panels and discussions with the scientific community, about which programs are most important for developing and sustaining the solar and space physics enterprise.

This report has been reviewed in draft form by individuals chosen for their diverse perspectives and technical expertise, in accordance with procedures approved by the National Research Council's Report Review Committee. The purpose of this independent review is to provide candid and

---

[2]The recommendations of each panel can be found in the companion volume to this report, *The Sun to the Earth—and Beyond: Panel Reports*, 2003, in press.

critical comments that will assist the institution in making its published report as sound as possible and to ensure that the report meets institutional standards for objectivity, evidence, and responsiveness to the study charge. The review comments and draft manuscript remain confidential to protect the integrity of the deliberative process. We wish to thank the following individuals for their review of this report:

> Claudia Alexander, California Institute of Technology,
> Lewis Allen, California Institute of Technology (retired),
> George Field, Harvard University,
> Peter Gilman, National Center for Atmospheric Research,
> Gerhard Haerendel, International University, Bremen, Germany,
> Thomas Hill, Rice University,
> W. Jeffrey Hughes, Boston University,
> Ralph Jacobson, The Charles Stark Draper Laboratory (retired),
> Robert Lin, University of California, Berkeley,
> Nelson Maynard, Mission Research Corporation,
> Atsuhiro Nishida, Japan Society for the Promotion of Science,
> William Radasky, Metatech Corporation, and
> Donald Williams, Johns Hopkins University Applied Physics Laboratory.

Although the reviewers listed above have provided many constructive comments and suggestions, they were not asked to endorse the conclusions or recommendations, nor did they see the final draft of the report before its release. The review of this report was overseen by Robert A. Frosch, Harvard University, and Lennard Fisk, University of Michigan. Appointed by the National Research Council, they were responsible for making certain that an independent examination of this report was carried out in accordance with institutional procedures and that all review comments were carefully considered. Responsibility for the final content of this report rests entirely with the authoring committee and the institution.

Louis J. Lanzerotti, *Chair*
Solar and Space Physics Survey Committee

Contents

Executive Summary   1

1   Solar and Space Physics: Milestones and Science Challenges   22
   The Domain of Solar and Space Physics, 23
   Milestones: From Stonehenge to SOHO, 31
   Science Challenges, 41
   The Astrophysical Context, 49
   Understanding Complex, Coupled Systems, 50
   Notes, 50

2   Integrated Research Strategy for Solar and Space Physics   53
   The Sun's Dynamic Interior and Corona, 54
   The Heliosphere and Its Components, 57
   Space Environments of Earth and Other Solar System Bodies, 58
   The Role of Theory and Modeling in Missions and Fundamental
      Space Plasma Physics, 64
   Space Weather, 66
   Roadmap to Understanding, 68
   Deferred High-Priority Flight Missions, 78
   Summary, 78
   Notes, 80

3   Technology Development   81
   Traveling to the Planets and Beyond, 83
   Advanced Spacecraft Systems, 85
   Advanced Science Instrumentation, 86
   Gathering and Assimilating Data from Multiple Platforms, 88
   Modeling the Space Environment, 89

Observing Geospace from Earth, 90
Observing the Magnetic Sun at High Resolution, 91
Notes, 92

4 Connections Between Solar and Space Physics and Other Disciplines 93
Laboratory Plasma Physics, 94
Astrophysical Plasmas, 98
Atmospheric Science and Climatology, 104
Atomic and Molecular Physics and Chemistry, 108
Notes, 109

5 Effects of the Solar and Space Environment on Technology and Society 111
Challenges Posed by Earth's Space Environment, 111
The National Space Weather Program, 115
Monitoring the Solar-Terrestrial Environment, 117
The Transition from Research to Operations, 120
Data Acquisition and Availability, 122
The Public and Private Sectors in Space Weather Applications, 124
Notes, 125

6 Education and Public Outreach 126
Educating Future Solar and Space Physicists, 127
Enhancing Education in Science and Technology, 136
Notes, 145

7 Strengthening the Solar and Space Physics Research Enterprise 147
A Strengthened Research Community, 147
Cost-Effective Use of Existing Resources, 150
Access to Space, 151
Interagency Cooperation and Coordination, 158
Facilitating International Partnerships, 159
Notes, 161

Appendixes

A	Statement of Task	165
B	Acronyms and Abbreviations	168
C	Biographical Information for Members of the Solar and Space Physics Survey Committee	171

Executive Summary

SCIENCE CHALLENGES

The Sun is the source of energy for life on Earth and is the strongest modulator of the human physical environment. In fact, the Sun's influence extends throughout the solar system, both through photons, which provide heat, light, and ionization, and through the continuous outflow of a magnetized, supersonic ionized gas known as the solar wind. The realm of the solar wind, which includes the entire solar system, is called the heliosphere. In the broadest sense, the heliosphere is a vast interconnected system of fast-moving structures, streams, and shock waves that encounter a great variety of planetary and small-body surfaces, atmospheres, and magnetic fields. Somewhere far beyond the orbit of Pluto, the solar wind is finally stopped by its interaction with the interstellar medium, which produces a termination shock wave and, finally, the outer boundary of the heliosphere. This distant region is the final frontier of solar and space physics.

During the 1990s, space physicists peered inside the Sun with Doppler imaging techniques to obtain the first glimpses of mechanisms responsible for the solar magnetic dynamo. Further, they imaged the solar atmosphere from visible to x-ray wavelengths to expose dramatically the complex interaction between the ionized gas and the magnetic field, which drives both the solar wind and energetic solar events such as flares and coronal mass ejections that strongly affect Earth. An 8-year tour of Jupiter's magnetosphere, combined with imaging from the Hubble Space Telescope, has revealed completely new phenomena resident in a regime dominated by planetary rotation, volcanic sources of charged particles, mysteriously pulsating x-ray auroras, and even an embedded satellite magnetosphere.

The response of Earth's magnetosphere to variations in the solar wind was clearly revealed by an international flotilla of more than a dozen spacecraft and by the first neutral-atom and extreme-ultraviolet imaging of energetic particles and cold plasma. At the same time, computer models of the global dynamics of the magnetosphere and of the local microphysics of magnetic reconnection have reached a level of sophistication high enough to enable verifiable predictions.

While the accomplishments of the past decades have answered important questions about the physics of the Sun, the interplanetary medium, and the space environments of Earth and other solar system bodies, they have also highlighted other questions, some of which are long-standing and fundamental. This report organizes these questions in terms of five challenges that are expected to be the focus of scientific investigations in solar and space physics during the coming decade and beyond:

- *Challenge 1: Understanding the structure and dynamics of the Sun's interior, the generation of solar magnetic fields, the origin of the solar cycle, the causes of solar activity, and the structure and dynamics of the corona.* Why does solar activity vary in a regular 11-year cycle? Why is the solar corona several hundred times hotter than its underlying visible surface, and how is the supersonic solar wind produced?

- *Challenge 2: Understanding heliospheric structure, the distribution of magnetic fields and matter throughout the solar system, and the interaction of the solar atmosphere with the local interstellar medium.* What is the nature of the interstellar medium, and how does the heliosphere interact with it? How do energetic solar events propagate through the heliosphere?

- *Challenge 3: Understanding the space environments of Earth and other solar system bodies and their dynamical response to external and internal influences.* How does Earth's global space environment respond to solar variations? What are the roles of planetary ionospheres, planetary rotation, and internal plasma sources in the transfer of energy among planetary ionospheres and magnetospheres and the solar wind?

- *Challenge 4: Understanding the basic physical principles manifest in processes observed in solar and space plasmas.* How is magnetic field energy converted to heat and particle kinetic energy in magnetic reconnection events?

- *Challenge 5: Developing a near-real-time predictive capability for understanding and quantifying the impact on human activities of dynamical processes at the Sun, in the interplanetary medium, and in Earth's magnetosphere and ionosphere.* What is the probability that specific types of space weather phenomena will occur over periods from hours to days?

An effective response to these challenges will require a carefully crafted program of space- and ground-based observations combined with, and guided by, comprehensive theory and modeling efforts. Success in this endeavor will depend on the ability to perform high-resolution imaging and in situ measurements of critical regions of the solar system. In addition to advanced scientific instrumentation, it will be necessary to have affordable constellations of spacecraft, advanced spacecraft power and propulsion systems, and advanced computational resources and techniques.

This report summarizes the state of knowledge about the total heliospheric system, poses key scientific questions for further research, and presents an integrated research strategy, with prioritized initiatives, for the next decade. The recommended strategy embraces both basic research programs and targeted basic research activities that will enhance knowledge and prediction of space weather effects on Earth. The report emphasizes the importance of understanding the Sun, the heliosphere, and planetary magnetospheres and ionospheres as astrophysical objects and as laboratories for the investigation of fundamental plasma physics phenomena. The recommendations presented in the main report are listed also in this Executive Summary.

AN INTEGRATED RESEARCH STRATEGY FOR SOLAR AND SPACE PHYSICS

The integrated research strategy proposed by the Solar and Space Physics Survey Committee is based on recommendations from four technical study panels regarding research initiatives in the following subject areas: solar and heliospheric physics, solar wind-magnetosphere interactions, atmosphere-ionosphere-magnetosphere interactions, and theory, computation, and data exploration. Because it was charged with recommending a program that will be feasible and responsible within a realistic resource envelope, the committee could not adopt all of the panels' recommendations. The committee's final set of recommended initiatives thus represents a prioritized selection from a larger set of initiatives recommended by the study panels. (All of the panel recommendations can be found in the second volume of this report, *The Sun to the Earth—and Beyond: Panel Reports*, 2003, in press.)

The committee organized the initiatives that it considered into four categories: large programs, moderate programs, small programs, and vitality programs. Moderate and small programs comprise both space missions and ground-based facilities and are defined according to cost, with moderate programs falling in the range from $250 million to $400 million and small programs costing less than $250 million. The committee considered one large (>$400 million) program, a Solar Probe mission, and gave it high priority for implementation in the decade 2003-2013. The programs in the vitality category are those that relate to the infrastructure for solar and space physics research; they are regarded by the committee as essential for the health and vigor of the field. The cost estimates used by the committee for all four categories are based either on the total mission cost or, for level-of-effort programs, on the total cost for the decade 2003-2013. FY 2002 costs are used in each case.

In arriving at a final recommended set of initiatives, the committee prioritized the selected initiatives according to two criteria—scientific importance and societal benefit. The ranked initiatives are listed and described briefly in Table ES.1. As discussed in Chapter 2, the rankings in Table ES.1, cost estimates, and judgments of technical readiness were then used to arrive at an overall program that could be conducted in the next decade while remaining within a reasonable budget. Nearly all of the recommended missions and facilities either are already planned or were recommended in previous strategic planning exercises conducted by the National Aeronautics and Space Administration (NASA) and the National Science Foundation (NSF).

The committee's recommended phasing of NASA missions and initiatives is shown in Figures ES.1 and ES.2; its recommended phasing of NSF initiatives is shown in Figure ES.3. While the committee did not find a need to create completely new mission or facility concepts, some existing programs are recommended for revitalization and will require stepwise or ramped funding increases. These programs include NASA's Suborbital Program, its Supporting Research and Technology (SR&T) program, and the University-Class Explorer (UNEX) program, as well as guest investigator initiatives in the NSF for national facilities. In the vitality category, new theory and modeling initiatives, notably the Coupling Complexity initiative (discussed in the report of the Panel on Theory, Modeling, and Data Exploration) and the Virtual Sun initiative (discussed in the report of the Panel on the Sun and Heliospheric Physics), are recommended.

Recommendation: The committee recommends the approval and funding of the prioritized programs listed in Table ES.1.

The committee developed its national strategy based on a systems approach to understanding the physics of the coupled solar-heliospheric environment. Ongoing NSF programs and facilities in solar and space physics, two complementary mission lines in the NASA Sun-Earth Connection program—the Solar Terrestrial Probes (STP) for basic research and Living With a Star (LWS) for targeted basic research—and applications and operations activities in the National Oceanic and Atmospheric Administration (NOAA) and the Department of Defense (DOD) facilitate such an approach.

As a key first element of its systems-oriented strategy, the committee endorsed three approved NASA missions: Solar-B and the Solar Terrestrial Relations Observatory (STEREO), both part of STP, and the Solar Dynamics Observatory (SDO), part of LWS. Together with ongoing NSF-supported solar physics programs and facilities as well as the start of the Advanced Technology Solar Telescope (ATST), these missions constitute a synergistic approach to the study of the inner heliosphere that will involve coordinated observations of the solar interior and atmosphere and the formation, release, evolution, and propagation of coronal mass ejections toward Earth. Later in the decade covered by the survey, overlapping investigations by the SDO, the ATST, and Magnetospheric Multiscale (MMS) (part of STP), together with the start of the Frequency-Agile Solar Radiotelescope (FASR), will form the intellectual basis for a comprehensive study of magnetic reconnection in the dense plasma of the solar atmosphere and the tenuous plasmas of geospace.

The committee's ranking of the Geospace Electrodynamic Connections (GEC; STP) and Geospace Network (LWS) missions acknowledges the importance of studying Earth's ionosphere and inner magnetosphere as a coupled system. Together with a ramping up of the launch opportunities in the Suborbital Program and the implementation of both the Advanced Modular Incoherent Scatter Radar (AMISR) and the Small Instrument Distributed Ground-Based Network, these missions will provide a unique opportunity to study the local electrodynamics of the ionosphere down to altitudes where energy is transferred between the magnetosphere and the atmosphere, while simultaneously investigating the global dynamics of the ionosphere and radiation belts. The implementation of the L1 Monitor (NOAA) and of the vitality programs will be essential to the success of this systems approach to basic and targeted basic research. Later on in the committee's recommended program, concurrent operations of a Multi-spacecraft Heliospheric Mission (MHM; LWS), Stereo Magnetospheric Imager (SMI; STP), and Magnetospheric Constellation (MagCon; STP) will provide opportunities for a coordinated approach to understanding the large-scale dynamics of the inner heliosphere and Earth's magnetosphere (again with strong contributions from the ongoing and new NSF initiatives).

TABLE ES.1 Priority Order of the Recommended Programs in Solar and Space Physics

Type of Program	Rank	Program	Description
Large	1	Solar Probe	Spacecraft to study the heating and acceleration of the solar wind through in situ measurements and some remote-sensing observations during one or more passes through the innermost region of the heliosphere (from ~0.3 AU to as close as 3 solar radii above the Sun's surface).
Moderate	1	Magnetospheric Multiscale	Four-spacecraft cluster to investigate magnetic reconnection, particle acceleration, and turbulence in magnetospheric boundary regions.
	2	Geospace Network	Two radiation-belt-mapping spacecraft and two ionospheric mapping spacecraft to determine the global response of geospace to solar storms.
	3	Jupiter Polar Mission	Polar-orbiting spacecraft to image the aurora, determine the electrodynamic properties of the Io flux tube, and identify magnetosphere-ionosphere coupling processes.
	4	Multispacecraft Heliospheric Mission	Four or more spacecraft with large separations in the ecliptic plane to determine the spatial structure and temporal evolution of coronal mass ejections (CMEs) and other solar-wind disturbances in the inner heliosphere.
	5	Geospace Electrodynamic Connections	Three to four spacecraft with propulsion for low-altitude excursions to investigate the coupling among the magnetosphere, the ionosphere, and the upper atmosphere.
	6	Suborbital Program	Sounding rockets, balloons, and aircraft to perform targeted studies of solar and space physics phenomena with advanced instrumentation.
	7	Magnetospheric Constellation	Fifty to a hundred nanosatellites to create dynamic images of magnetic fields and charged particles in the near magnetic tail of Earth.
	8	Solar Wind Sentinels	Three spacecraft with solar sails positioned at 0.98 AU to provide earlier warning than L1 monitors and to measure the spatial and temporal structure of CMEs, shocks, and solar-wind streams.
	9	Stereo Magnetospheric Imager	Two spacecraft providing stereo imaging of the plasmasphere, ring current, and radiation belts, along with multispectral imaging of the aurora.
Small	1	Frequency-Agile Solar Radiotelescope	Wide-frequency-range (0.3-30 GHz) radiotelescope for imaging of solar features from a few hundred kilometers above the visible surface to high in the corona.

TABLE ES.1 Continued

Type of Program	Rank	Program	Description
	2	Advanced Modular Incoherent Scatter Radar	Movable incoherent scatter radar with supporting optical and other ground-based instruments for continuous measurements of magnetosphere-ionosphere interactions.
	3	L1 Monitor	Continuation of solar-wind and interplanetary magnetic field monitoring for support of Earth-orbiting space physics missions. Recommended for implementation by NOAA.
	4	Solar Orbiter	U.S. instrument contributions to European Space Agency spacecraft that periodically corotates with the Sun at 45 solar radii to investigate the magnetic structure and evolution of the solar corona.
	5	Small Instrument Distributed Ground-Based Network	NSF program to provide global-scale ionospheric and upper atmospheric measurements for input to global physics-based models.
	6	University-Class Explorer	Revitalization of University-Class Explorer program for more frequent access to space for focused research projects.
Vitality	1	NASA Supporting Research and Technology	NASA research and analysis program.
	2	National Space Weather Program	Multiagency program led by the NSF to support focused activities that will improve scientific understanding of geospace in order to provide better specifications and predictions.
	3	Coupling Complexity	NASA/NSF theory and modeling program to address multiprocess coupling, nonlinearity, and multiscale and multiregional feedback.
	4	Solar and Space Physics Information System	Multiagency program for integration of multiple data sets and models in a system accessible by the entire solar and space physics community.
	5	Guest Investigator Program	NASA program for broadening the participation of solar and space physicists in space missions.
	6	Sun-Earth Connection Theory and LWS Data Analysis, Theory, and Modeling Programs	NASA programs to provide long-term support to critical-mass groups involved in specific areas of basic and targeted basic research.
	7	Virtual Sun	Multiagency program to provide a systems-oriented approach to theory, modeling, and simulation that will ultimately provide continuous models from the solar interior to the outer heliosphere.

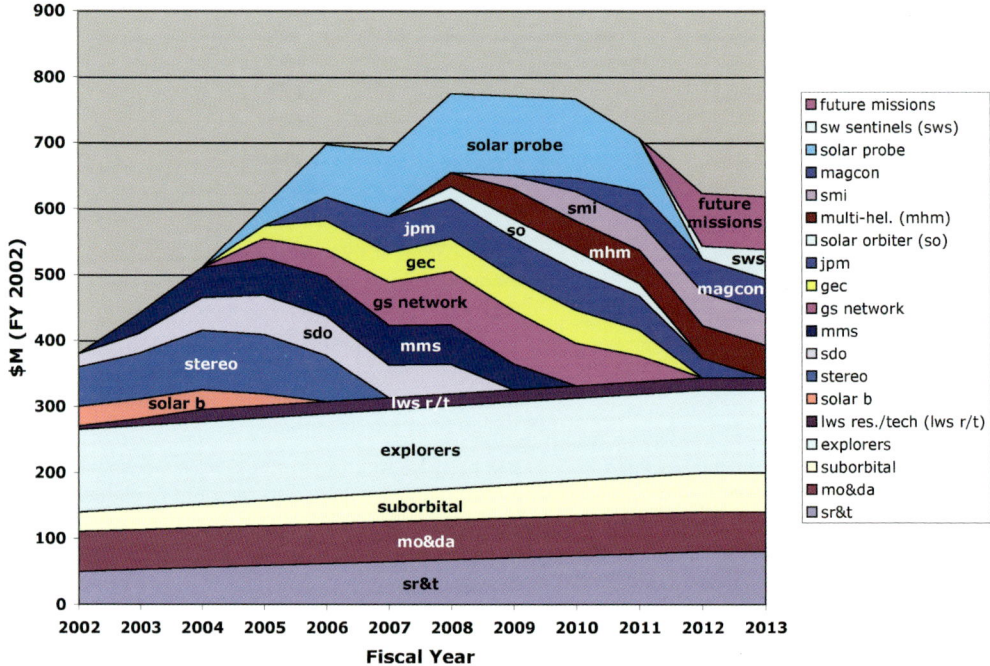

FIGURE ES.1 Recommended phasing of the highest-priority NASA missions, assuming an early implementation of a Solar Probe mission. Solar Probe was the Survey Committee's highest priority in the large mission category, and the committee recommends its implementation as soon as possible. However, the projected cost of Solar Probe is too high to fit within plausible budget and mission profiles for NASA's Sun-Earth Connection (SEC) Division. Thus, as shown in this figure, an early start for Solar Probe would require funding above the currently estimated SEC budget of $650 million per year for fiscal years 2006 and beyond. Note that mission operations and data analysis (MO&DA) costs for all missions are included in the MO&DA budget wedge.

To understand the genesis of the heliospheric system, it is necessary to determine the mechanisms by which the solar corona is heated and the solar wind is accelerated and to understand how the solar wind evolves in the innermost heliosphere. These objectives will be addressed by a Solar Probe mission. Because of the importance of these objectives for the overall understanding of the solar-heliosphere system, as well as of other stellar systems, a Solar Probe mission[1] should be implemented as soon as possible within the coming decade. The Solar Probe measurements will be comple-

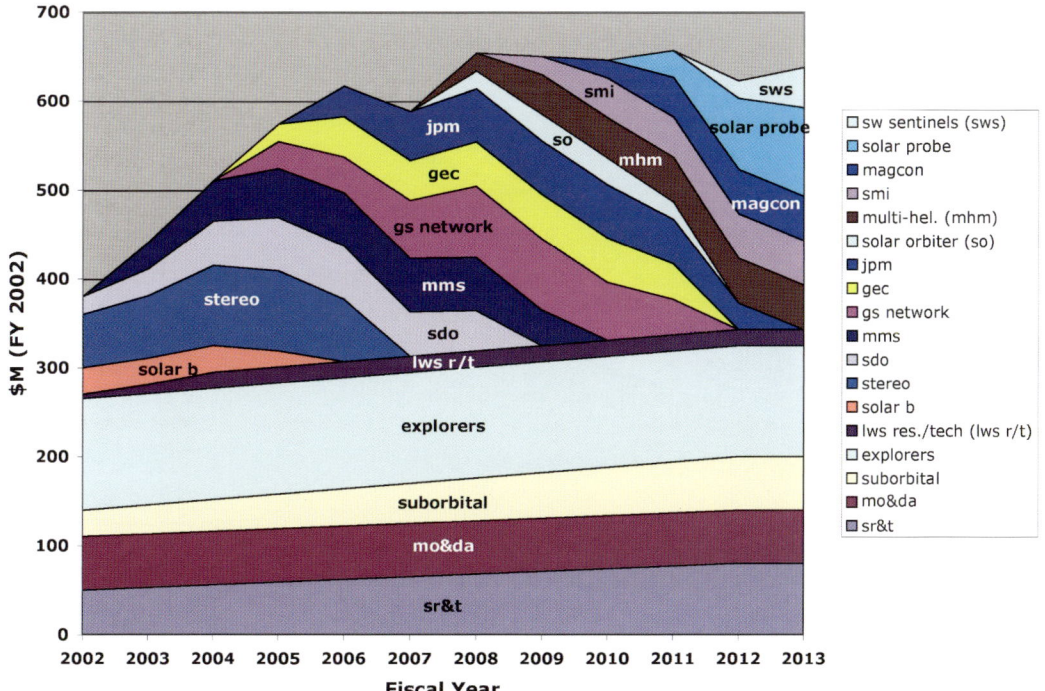

FIGURE ES.2 Recommended phasing of the highest-priority NASA missions if budget augmentation for Solar Probe is not obtained. MO&DA costs for all missions are included in the MO&DA budget wedge.

mented by correlative observations from such initiatives as Solar Orbiter, SDO, ATST, and FASR.

Similarly, because comparative magnetospheric studies are important for advancing the understanding of basic magnetospheric processes, the committee has assigned high priority to a Jupiter Polar Mission (JPM), a space physics mission to study high-latitude electrodynamic coupling at Jupiter. Such a mission will provide both a means of testing and refining theoretical concepts developed largely in studies of the terrestrial magnetosphere and a means of studying in situ the electromagnetic redistribution of angular momentum in a rapidly rotating system, with results relevant to such astrophysical questions as the formation of protostars.

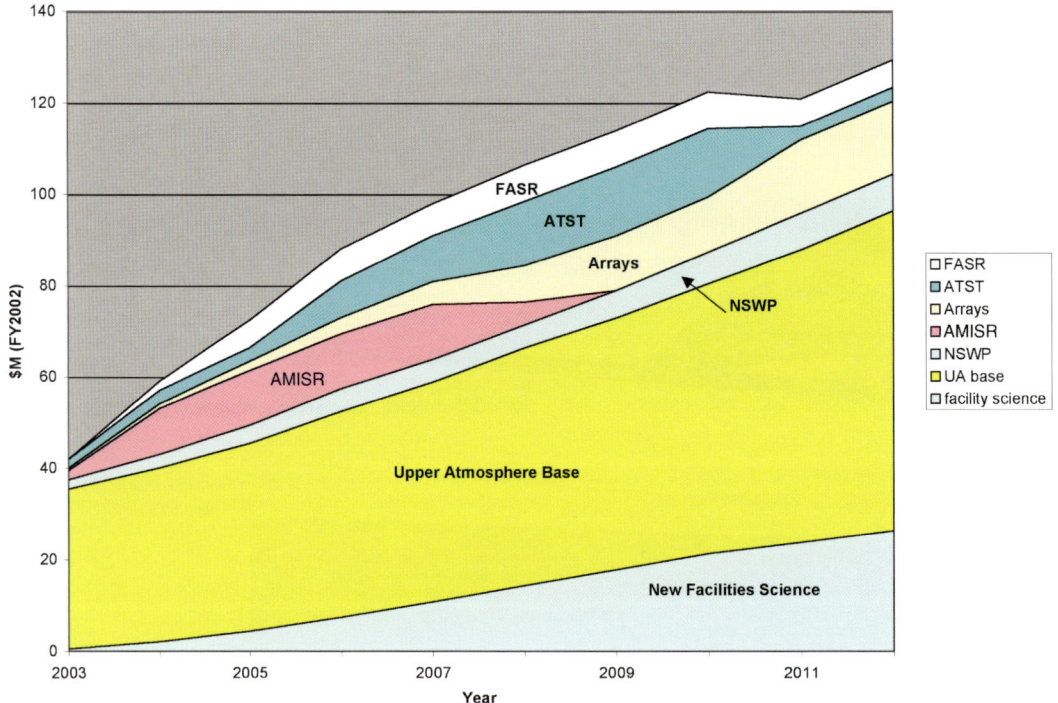

FIGURE ES.3 Recommended phasing of major new and enhanced NSF initiatives. The budget wedge for new facilities science refers to support for guest investigator and related programs that will maximize the science return of new ground facilities to the scientific community. Funding for new facilities science is budgeted at approximately 10 percent of the aggregate cost for new NSF facilities.

TECHNOLOGY DEVELOPMENT

Technology development is required in several critical areas if a number of the future science objectives of solar and space physics are to be accomplished.

Traveling to the planets and beyond. New propulsion technologies are needed to rapidly propel spacecraft to the outer fringes of the solar system and into the local interstellar medium. Also needed are power systems to support future deep-space missions.

Recommendation: NASA should assign high priority to the development of advanced propulsion and power technologies required for the

exploration of the outer planets, the inner and outer heliosphere, and the local interstellar medium. Such technologies include solar sails, space nuclear power systems, and high-efficiency solar arrays. Equally high priority should be given to the development of lower-cost launch vehicles for Explorer-class missions and to the reopening of the radioisotope thermoelectric generator (RTG) production line.

Advanced spacecraft systems. Highly miniaturized spacecraft and advanced spacecraft subsystems will be critical for a number of high-priority future missions and programs in solar and space physics.

Recommendation: NASA should continue to give high priority to the development and testing of advanced spacecraft technologies through initiatives such as the New Millennium Program and its advanced technology program.

Advanced science instrumentation. Highly miniaturized sensors of charged and neutral particles and photons will be essential elements of instruments for new solar and space physics missions.

Recommendation: NASA should continue to assign high priority, through its recently established new instrument development programs, to supporting the development of advanced instrumentation for solar and space physics missions and programs.

Gathering and assimilating data from multiple platforms. Future flight missions include multipoint measurements to resolve spatial and temporal scales that dominate the physical processes that operate in solar system plasmas.

Recommendation: NASA should accelerate the development of command-and-control and data acquisition technologies for constellation missions.

Modeling the space environment. Primarily because of the lack of a sufficient number of measurements, it has not been necessary until quite recently for the solar and space physics community to address data assimilation issues. However, it is anticipated that within 10 years vast arrays of data sets will be available for assimilation into models.

Recommendation: Existing NOAA and DOD facilities should be expanded to accommodate the large-scale integration of space- and ground-based data sets into physics-based models of the geospace environment.

Observing geospace from Earth. The effects of temperature, moisture, and wildly varying solar insolation have posed serious problems for arrays of ground-based sensor systems that are critical for solar and space physics studies.

Recommendation: The relevant program offices in the NSF should support comprehensive new approaches to the design and maintenance of ground-based, distributed instrument networks, with proper regard for the severe environments in which they must operate.

Observing the Sun at high spatial resolution. Recent breakthroughs in adaptive optics have eliminated the major technical impediments to making solar observations with sufficient resolution to measure the pressure scale height, the photon mean free path, and the fundamental magnetic structure size.

Recommendation: The NSF should continue to fund the technology development program for the Advanced Technology Solar Telescope.

CONNECTIONS BETWEEN SOLAR AND SPACE PHYSICS AND OTHER DISCIPLINES

The fully or partially ionized plasmas that are the central focus of solar and space physics are related on a fundamental level to laboratory plasma physics, which directly investigates basic plasma physical processes, and to astrophysics, a discipline that relies heavily on understanding the physics unique to the plasma state. Moreover, there are numerous points of contact between space physics and atmospheric science, particularly in the area of aeronomy. Knowledge of the properties of atoms and molecules is critical for understanding a number of magnetospheric, ionospheric, solar, and heliospheric processes. Understanding developed in one of these fields is thus in principle applicable to the others, and productive cross-fertilization between disciplines has occurred in a number of instances.

Recommendation: In collaboration with other interested agencies, the NSF and NASA should take the lead in initiating a program in laboratory plasma science that can provide new understanding of fundamental processes important to solar and space physics.[2]

Recommendation: The NSF and NASA should take the lead and other interested agencies should collaborate in supporting, via the proposal and funding processes, increased interactions between researchers in

solar and space physics and those in allied fields such as atomic and molecular physics, laboratory fusion physics, atmospheric science, and astrophysics.

SOLAR AND SPACE ENVIRONMENT EFFECTS ON TECHNOLOGY AND SOCIETY

The space environment of the Sun-Earth system can have deleterious effects on numerous technologies that are used by modern-day society. Understanding this environment is essential for the successful design, implementation, and operation of these technologies.

National Space Weather Program. A number of activities under way in the United States aim to better understand and to mitigate the effects of solar activity and the space environment on important technological systems. The mid-1990s saw the creation of the National Space Weather Program (NSWP), an interagency effort whose goal is to achieve, within a 10-year period, "an active, synergistic, interagency system to provide timely, accurate, and reliable space environment observations, specifications, and forecasts."[3] In 1999, NASA initiated an important complementary program, Living With a Star, which over the next decade and beyond will carry out targeted basic research on space weather. Crucial components of the national space weather effort continue to be provided by the operational programs of the Department of Defense and NOAA. Moreover, in addition to governmental activities, a number of private companies have, over the last decade, become involved in developing and providing space weather products.

Monitoring the solar-terrestrial environment. Numerous research instruments and observations are required to provide the basis for modeling interactions between the solar-terrestrial environment and technical systems and for making sound technical design decisions that take such interactions into account. Transitioning of programs and/or their acquisition platforms or instruments into operational use requires strong and effective coordination efforts among agencies. Imaging of the Sun and of geospace will play a central role in operational space forecasting in the future.

Recommendation: NOAA and DOD, in consultation with the research community, should lead in an effort by all involved agencies to jointly assess instrument facilities that contribute key data to public

and private space weather models and to operational programs. They should then determine a strategy to maintain the needed facilities and/or work to establish new facilities. The results of this effort should be available for public dissemination.

Recommendation: NOAA should assume responsibility for the continuance of space-based measurements such as solar wind data from the L1 location as well as near Earth and for distribution of the data for operational use.[4]

Recommendation: NASA and NOAA should initiate the necessary planning to transition solar and geospace imaging instrumentation into operational programs for the public and private sectors.

Transition from research to operations. Means must be established for transitioning new knowledge into those arenas where it is needed for design and operational purposes. Creative and cutting-edge research in modeling the solar-terrestrial environment is under way. Under the auspices of the NSWP, models that are thought to be potentially useful for space weather applications can be submitted to the Community Coordinated Modeling Center (currently located at the NASA Goddard Space Flight Center) for testing and validation. Following validation, the models can be turned over to either the U.S. Air Force or the NOAA Rapid Prototyping Center, where the models are used for the objectives of the individual agencies. In many instances, the validation of research products and models is different in the private and public sectors, with publicly funded research models and system-impact products usually being placed in an operational setting with only limited validation.

Recommendation: The relevant federal agencies should establish an overall verification and validation program for all publicly funded models and system-impact products before they become operational.

Recommendation: The operational federal agencies, NOAA and DOD, should establish procedures to identify and prioritize operational needs, and these needs should determine which model types are selected for transitioning by the Community Coordinated Modeling Center and the Rapid Prototyping Centers. After the needs have been prioritized, procedures should be established to determine which of the competing models, public or private, is best suited for a particular operational requirement.

Data acquisition and availability. During the coming decade, gigabytes of data could be available every day for incorporation into physics-based data assimilation models of the solar-terrestrial environment and into system-impact codes for space weather forecasting and mitigation purposes. DOD generally uses data that it owns and only recently has begun to use data from other agencies and institutions, so that not many data sets are available for use by the publicly funded or commercial vendors who design products for DOD. Engineers typically are interested in space climate, not space weather. Needed are long-term averages, the uncertainties in these averages, and values for the extremes in key space weather parameters. The engineering goal is to design systems that are as resistant as possible to the effects of space weather.

Recommendation: DOD and NOAA should be the lead agencies in acquiring all the data sets needed for accurate specification and forecast modeling, including data from the international community. Because it is extremely important to have real-time data, both space- and ground-based, for predictive purposes, NOAA and DOD should invest in new ways to acquire real-time data from all of the ground- and space-based sources available to them. All data acquired should contain error estimates, which are required by data assimilation models.

Recommendation: A new, centralized database of extreme space weather conditions should be created that covers as many of the relevant space weather parameters as possible.

Public and private sectors in space weather applications. To date, the largest efforts to understand the solar-terrestrial environment and apply the resultant gains in knowledge for practical purposes have been mostly publicly funded and have involved government research organizations, universities, and some industries. Recently some private companies both large and small have been devoting their own resources to the development and sale of specialized products that address the design and operation of certain technical systems that can be affected by the solar-terrestrial environment. Such companies often use publicly supported assets (such as spacecraft data) as well as proprietary instrumentation and models. A number of the private efforts use proprietary system knowledge to guide their choice of research directions. Policies on such matters as data rights, intellectual property rights and responsibilities, and benchmarking criteria can be quite different for private efforts and publicly supported ones, including those of universities. Thus, transitioning knowledge and models from one sector to another can be fraught with complications and requires continued attention and discussion by all interested entities.

Recommendation: Clear policies should be developed that describe government and industry roles, rights, and responsibilities in space weather activities. Such policies are necessary to optimize the benefits of the national investments, public and private, that are being made.

EDUCATION AND PUBLIC OUTREACH

The committee's consideration of issues related to education and outreach was focused in two areas: ensuring a sufficient number of future scientists in solar and space physics and identifying ways in which the solar and space physics community can contribute to national initiatives in science and technology education.

Solar and space physics in colleges and universities. Because of its relatively short history, solar and space physics appears only adventitiously in formal instructional programs, and an appreciation of its importance is often lacking in current undergraduate curricula. If solar and space physics is to have a healthy presence in academia, additional faculty members will be needed to guide student research (both undergraduate and graduate), to teach solar and space physics graduate programs, and to integrate topics in solar and space physics into basic physics and astronomy classes.

Recommendation: The NSF and NASA should jointly establish a program of "bridged positions" that provides (through a competitive process) partial salary, start-up funding, and research support for four new faculty members every year for 5 years.

Distance education. Education in solar and space physics during the academic year could be considerably enhanced if the latest advances in information technology are exploited to provide distance learning for both graduate students and postdoctoral researchers. This approach would substantially increase the educational value of the expertise that currently resides at a limited number of institutions.

Recommendation: The NSF and NASA should jointly support an initiative that provides increased opportunities for distance education in solar and space physics.

Undergraduate research opportunities and undergraduate instruction. NSF support for the Research Experiences for Undergraduates program has been valuable for encouraging undergraduates in the solar and space physics research area.

> **Recommendation: NASA should institute a specific program for the support of undergraduate research in solar and space physics at colleges and universities. The program should have the flexibility to support such research with either a supplement to existing grants or with a stand-alone grant.**

> **Recommendation: Over the next decade NASA and the NSF should fund groups to develop and disseminate solar and space physics educational resources (especially at the undergraduate level) and to train educators and scientists in the effective use of such resources.**

STRENGTHENING THE SOLAR AND SPACE PHYSICS RESEARCH ENTERPRISE

Advances in understanding in solar and space physics will require strengthening a number of the infrastructural aspects of the nation's solar and space physics program. The committee has identified several that depend on effective program management and policy actions for their success: (1) development of a stronger research community, (2) cost-effective use of existing resources, (3) ensuring cost-effective and reliable access to space, (4) improving interagency cooperation and coordination, and (5) facilitating international partnerships.

Strengthening the solar and space physics research community. A diverse and high-quality community of research institutions has contributed to solar and space physics research over the years. The central role of the universities as research sites requires enhancement, strengthening, and stability.

> **Recommendation: NASA should undertake an independent outside review of its existing policies and approaches regarding the support of solar and space physics research in academic institutions, with the objective of enabling the nation's colleges and universities to be stronger contributors to this research field.**

Recommendation: NSF-funded national facilities for solar and space physics research should have resources allocated so that the facilities can be made widely available to outside users.

Cost-effective use of existing resources. Optimal return in solar and space physics is obtained not only through the judicious funding and management of new assets, but also through the maintenance and upgrading, funding, and management of existing facilities.

Recommendation: The NSF and NASA should give all possible consideration to capitalizing on existing ground- and space-based assets as the goals of new research programs are defined.

Access to space. The continuing vitality of the nation's space research program is strongly dependent on having cost-effective, reliable, and readily available access to space that meets the requirements of a broad spectrum of diverse missions. The solar and space physics research community is especially dependent on the availability of a wide range of suborbital and orbital flight capabilities to carry out cutting-edge science programs, to validate new instruments, and to train new scientists. Suborbital flight opportunities are very important for advancing many key aspects of future solar and space physics research objectives and for enabling the contributions that such opportunities make to education.

Recommendation: NASA should revitalize the Suborbital Program to bring flight opportunities back to previous levels.

Low-cost launch vehicles with a wide spectrum of capabilities are critically important for the next generation of solar and space physics research, as delineated in this report.

Recommendations:

• **NASA should aggressively support the engineering research and development of a range of low-cost vehicles capable of launching payloads for scientific research.**
• **NASA should develop a memorandum of understanding with DOD that would delineate a formal procedure for identifying in advance flights of opportunity for civilian spacecraft as secondary payloads on certain Air Force missions.**
• **NASA should explore the feasibility of similar piggybacking on appropriate foreign scientific launches.**

The comparative study of planetary ionospheres and magnetospheres is a central theme of solar and space physics research.

Recommendation: The scientific objectives of the NASA Discovery program should be expanded to include those frontier space plasma physics research subjects that cannot be accommodated by other spacecraft opportunities.

The principal investigator (PI) model that has been used for numerous Explorer missions has been highly successful. Strategic missions such as those under consideration for the STP and LWS programs can benefit from emulating some of the management approach and structure of the Explorer missions. The committee believes that the science objectives of the solar and space physics missions currently under consideration are best achieved through a PI mode of mission management.

Recommendation: NASA should (1) place as much responsibility as possible in the hands of the principal investigator, (2) define the mission rules clearly at the beginning, and (3) establish levels of responsibility and mission rules within NASA that are tailored to the particular mission and to its scope and complexity.

Recommendation: The NASA official who is designated as the program manager for a given project should be the sole NASA contact for the principal investigator. One important task of the NASA official would be to ensure that rules applicable to large-scale, complex programs are not being inappropriately applied, thereby producing cost growth for small programs.

Interagency cooperation and coordination. Interagency coordination over the years has yielded greater science returns than could be expected from single-agency activities. In the future, a research initiative at one agency could trigger a window of opportunity for a research initiative at another agency. Such an eventuality would leverage the resources contributed by each agency.

Recommendation: The principal agencies involved in solar and space physics research—NASA, NSF, NOAA, and DOD—should devise and implement a management process that will ensure a high level of coordination in the field and that will disseminate the results of such a coordinated effort—including data, research opportunities, and related matters—widely and frequently to the research community.

Recommendation: For space-weather-related applications, increased attention should be devoted to coordinating NASA, NOAA, NSF, and DOD research findings, models, and instrumentation so that new developments can quickly be incorporated into the operational and applications programs of NOAA and DOD.

International partnerships. The geophysical sciences—in particular, solar and space physics—address questions of global scope and inevitably require international participation for their success. Collaborative research with other nations allows the United States to obtain from other geographical regions data that are necessary to determine the global distributions of space processes. Studies in space weather cannot be successful without strong participation from colleagues in other countries and their research capabilities and assets, in space and on the ground.

Recommendation: Because of the importance of international collaboration in solar and space physics research, the federal government, especially the State Department and NASA, should implement clearly defined procedures regarding exchanges of scientific data or information on instrument characteristics that will facilitate the participation of researchers from universities, private companies, and nonprofit organizations in space research projects having an international component.

NOTES

1. The Solar Probe mission recommended by the committee is a generic mission to study the heating and acceleration of the solar wind through measurements as close to the surface of the Sun as possible. NASA's previously announced Solar Probe mission was canceled for budgetary reasons; a new concept study for a Solar Probe was conducted in 2002. The new study built on the earlier science definition team report to NASA and examined, among other issues, the power and communications technologies (including radioisotope thermoelectric generators) needed to enable such a mission within a realistic cost cap. The measurement capabilities considered in the study comprise both instrumentation for the in situ measurement of plasmas, magnetic fields, and waves and a remote-sensing package, including a magnetograph and Doppler, extreme ultraviolet, and coronal imaging instruments.

The committee notes that the Panel on the Sun and Heliospheric Physics recommends as its highest-priority new initiative a Solar Probe mission whose primary objective is to make in situ measurements of the innermost heliosphere. The panel does not consider remote sensing a top priority on a first mission to the near-Sun region, although it does allow as a possible secondary objective remote sensing of the photospheric magnetic field in the polar regions. (See the Solar Probe discussion in the report of the Panel on the Sun and Heliospheric Physics, which is published in *The Sun to the Earth—and Beyond: Panel Reports,* 2003, in press.) While accepting the panel's assessment of the critical importance of the in situ measurements for understanding coronal heating and solar wind acceleration, the committee does not wish to rule out the possibility that some additional remote-sensing capabilities, beyond the remote-sensing experiment to measure the polar photospheric magnetic field envisioned by the panel, can be accommodated on a Solar Probe within the cost cap set by the committee.

2. The establishment of such a laboratory initiative was previously recommended in the 1995 National Research Council report *Plasma Science: From Fundamental Research to Technological Applications* (National Academy Press, Washington, D.C., 1995).

3. Office of the Federal Coordinator for Meteorological Services and Supporting Research (OFCM), *National Space Weather Program Strategic Plan,* FCM-P30-1995, OFCM, Washington, D.C., August 1995.

4. For example, a NOAA-Air Force program is producing operational solar x-ray data. The Geostationary Operational Environmental Satellite (GOES) Solar X-ray Imager (SXI), first deployed on GOES-M, took its first image on September 7, 2001. The SXI instrument is designed to obtain a continuous sequence of coronal x-ray images at a 1-minute cadence. These images are being used by NOAA's Space Environment Center and the broader community to monitor solar activity for its effects on Earth's upper atmosphere and the near-space environment.

1
Solar and Space Physics: Milestones and Science Challenges

The fundamental goal of solar and space physics research is to discover, to explore, and ultimately to understand the activity of a star—the Sun—and the often complex effects of that activity on the interplanetary environment, the planets and other solar system bodies, and the interstellar medium. This enterprise involves the study of an exotic and dynamic world of ionized gases (plasmas), magnetic and electric fields, and small- and large-scale electrical currents. It is motivated by the deep-seated human impulse to know and understand the workings of Nature. Its province is that portion of the universe dominated by the Sun's activity. Although solar and space physicists focus their inquiry primarily on the behavior of magnetized plasmas in the solar system and on the interactions of these plasmas with each other and with electrically neutral matter, the processes that they seek to understand are fundamental, so that the lessons of solar and space physics are often relevant to our understanding of astrophysical objects lying well beyond the reach of the Sun's influence.

Intellectual inquiry into fundamental processes often yields utilitarian benefits of considerable value to society.[1] In the case of solar and space physics, utility is found principally in three areas. First, solar activity produces disturbances in Earth's space environment that can adversely affect certain important technologies and threaten the health and safety of astronauts. Knowledge obtained through solar and space physics research is essential to the development of means and strategies for mitigating the harmful effects of such disturbances. Second, global climate change is an issue of great scientific complexity and profound societal significance. Recent studies suggest that solar variability may have been responsible for a large fraction of the changes in global mean surface temperature that occurred prior to 1900 and that it continues to have a significant influence today. Understanding variations, both long- and short-term, in the Sun's magnetic activity and radiative output is one of the necessary conditions for distinguishing the human influence on global climate from the background of natural variability (see sidebar, "The Sun and Climate"). Third, as evidenced by the enthusiastic reception that audiences have given the documentary *SolarMax*,[2] the subject matter of solar and space physics, like that of their sister discipline, astronomy, exercises a powerful hold on the interest and imagination of the public. Solar physics and space physics, along with astronomy, are thus particularly well suited to contribute to the strengthening of science education and to the development of a scientifically literate public and a technically trained workforce.

THE DOMAIN OF SOLAR AND SPACE PHYSICS

The domain of solar and space physics is that region of our galaxy known as the heliosphere (Figure 1.1). The heliosphere is the cavity formed within the warm plasma of the local interstellar medium by the solar wind, the Sun's ionized, supersonically expanding atmosphere. Situated inside this bubblelike cavity, and immersed in the solar wind flow, are the nine planets of the solar system, the asteroids, the comets, and the icy trans-Neptunian objects of the Kuiper Belt. At its center is the Sun, an ordinary main-sequence star that, with an age of 4.5 billion years, has reached the midpoint of its stellar life. The boundaries of the heliosphere have not yet been surveyed because they are farther out than the most distant deep-space probe, Voyager 1, which in May 2002 was located some 86 astronomical units (AU)—more than 12.5 billion kilometers—from the Sun.[3]

The size and structure of the heliosphere are determined by the relative pressures of the solar wind and the interstellar medium and will change as these pressures vary. At this point in the history of the solar system, most of the changes in the dimensions of the heliosphere probably result from variations in solar wind ram pressure over the course of the 11-year solar cycle and are likely to be comparatively minor (a few percent). There is reason to believe, however, that the size of the heliosphere can vary dramatically with sufficiently large changes in the density of the local interstellar medium. Recent computer simulations show that an encounter with an interstellar cloud whose neutral hydrogen density is 50 times that of the heliosphere's present galactic environment could reduce the size of the heliosphere by as much as 80 percent, producing changes in the inner solar system that could affect Earth's space environment and climate.[4]

The wind that inflates the heliosphere, the solar wind, blows continuously. It originates in the several-million-degree solar corona and is accelerated to supersonic speeds near the Sun. Like its coronal source, the solar

THE SUN AND CLIMATE

The Sun is the ultimate driver of the climate system, and it is reasonable to suspect that there might be a link between solar variability and changes in climate. Despite many claims of correlations between solar-activity indicators and climate variables, the existence of such a link has been controversial. Part of the difficulty has revolved around the question of a physical mechanism to couple solar variability to the lower atmosphere and the surface of Earth.

Variations in irradiance are an obvious and plausible mechanism for solar influence on climate. The Sun's total irradiance (the solar "constant") varies on time scales at least as long as the 11-year solar activity cycle, but the variations directly observed have been too small (0.1 percent of the total irradiance) to have a significant impact. If larger variations occur, they can either weaken or amplify anthropogenic effects and thereby increase or reduce the time available to address these effects. It is of the utmost importance to estimate their magnitude.

Part of the variation in the Sun's irradiance takes place in the ultraviolet (UV) region of the spectrum that is responsible for both the formation and the destruction of ozone in the stratosphere. The relative UV variations are much larger than total luminosity variations and, as shown by recent modeling work, can have a significant effect on tropospheric dynamics and hence on climate.

Other, less direct mechanisms have been suggested whereby the Sun might influence climate, such as those in which solar-induced variability in cosmic-ray ionization affects cloud formation. Clouds have a major influence on climate, so if these links exist, they could provide a powerful means of amplifying the effects of solar variability. An alternative suggestion involves changes in the global electric field influencing the freezing of supercooled water droplets and the release of latent heat. Some observations have been quoted as support for these mechanisms, but they remain unsubstantiated and controversial, indicating the need for further study.

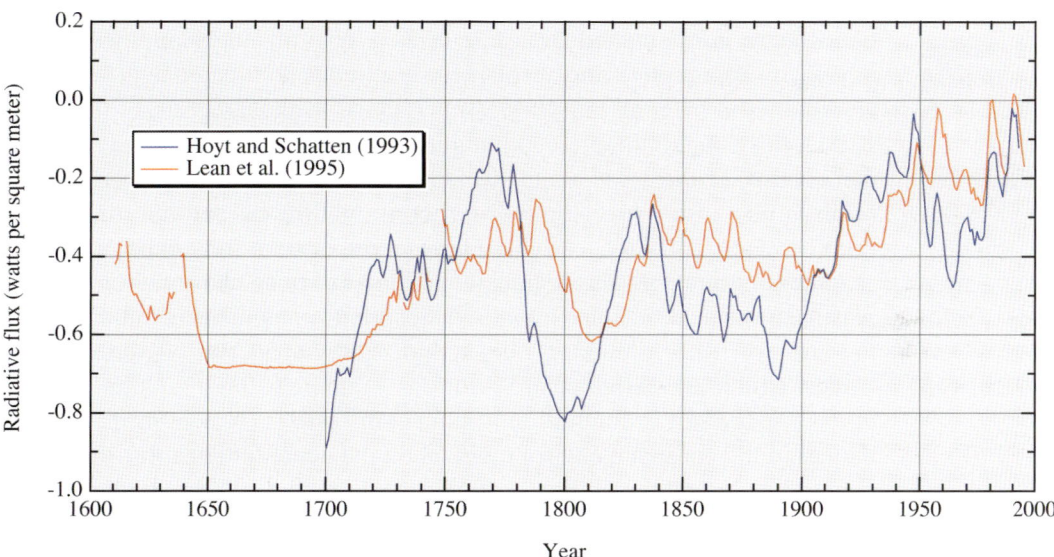

Two model-based estimates of the globally averaged solar radiative flux entering the atmosphere, shown as departures from the 1980 value of about 240 W m^{-2}. The Lean et al. estimate (Lean, J., J. Beer, and R. Bradley, Reconstruction of solar irradiance since 1610: Implications for climate change, *Geophysical Research Letters* 22, 3195-3198, 1995) is based on the known irradiance variation on the time scale of the 11-year solar magnetic cycle together with a long-term variation derived from cosmogenic isotope variations and the observed characteristics of sunlike stars. The Hoyt and Schatten estimate (Hoyt, D.V., and K.H. Schatten, A discussion of plausible solar irradiance variations, 1700-1992, *Journal of Geophysical Research* 98, 18895-18906, 1993) uses a long-term variation based on the length of the 11-year cycle rather than its amplitude, together with other parameters of solar variability. While there are significant differences between the two plots, their gross similarity can be taken to imply that the overall features of solar irradiance variability are known. A word of caution is necessary, however, since the mechanisms responsible for long-term irradiance variability are not well understood and could in principle involve as yet unknown changes in the transport of radiation through the Sun's convective zone. Courtesy of G.C. Reid.

wind is structured and variable. It varies in density, speed, and temperature, and in the strength and orientation of the magnetic field embedded in its flow (the interplanetary magnetic field, or IMF). At solar minimum, the heliosphere is dominated by a fast solar wind from high latitudes, while during the approach to and at solar maximum it is dominated by a slow and variable wind from all latitudes. This change in the structure of the solar wind reflects the dramatic reconfiguration of the corona that takes place as the polarity of the solar magnetic field reverses. During this period, the solar wind is also increasingly disturbed in its flow by coronal mass ejections (CMEs), which occur over ten times more often at solar maximum than at solar minimum. CMEs are transient releases of huge quantities of coronal plasma and magnetic fields into the heliosphere, sometimes at initial speeds in excess of 1,000 kilometers per second. Fast CMEs drive powerful shock waves, which accelerate solar wind ions to energies high enough to penetrate a space suit or the hull of a spacecraft[5] and can cause severe disturbances in the geospace environment when they encounter Earth's magnetic field. Developing the ability to predict CMEs is thus an important goal for solar and space physicists.

The eruption of CMEs and flares, the heating of the corona to temperatures several hundred times that of the Sun's visible surface, the acceleration of the solar wind—all of these processes, whose detailed workings are still poorly understood, are driven or mediated by energy provided by magnetic fields generated within the upper third of the Sun's interior, in the so-called convection zone. In this region, the rotational and turbulent convective motions of the electrically conducting plasma drive a magnetic dynamo that generates and maintains the Sun's global magnetic field as well as smaller-scale local fields. The magnetic fields thus generated emerge through the photosphere, forming sunspots and other active regions and creating the complex and dynamic coronal structures revealed in such stunning detail in recent images from the Transition Region and Coronal Explorer (TRACE) spacecraft (Figure 1.2). The last decade has seen considerable progress in theoretical and modeling studies of solar magnetism. However, important questions remain—for example, about the distribution, emergence, and evolution of magnetic flux on the Sun; about the storage and release of energy in solar magnetic fields; and about the detailed workings of the solar dynamo and the origins of the solar cycle. In particular, the astonishing fibril state of the magnetic field at the visible surface needs to be understood, along with the degree to which the field is in a fibril state far below the surface.

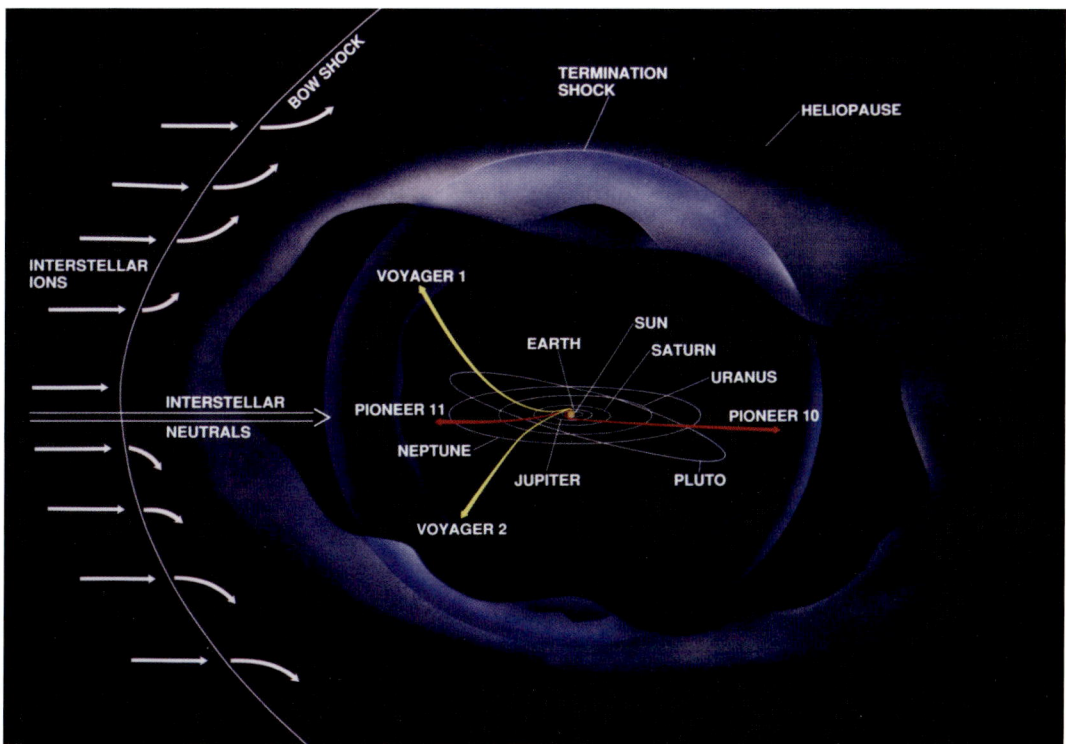

FIGURE 1.1 Artist's concept of the heliosphere, the cavity formed in the local interstellar medium by the solar wind. The dimensions of the heliosphere are not known, but its diameter is estimated to be on the order of 200 to 250 AU. A boundary referred to as the heliopause separates the solar wind from the magnetized interstellar plasma. Inside this boundary is a shock wave, the termination shock, where the speed of the solar wind changes from super- to subsonic. A shock wave may also form upstream of the heliosphere if the motion of the heliosphere through the interstellar medium is supersonic. Shown in the figure are the trajectories of the four deep-space probes—Pioneers 10 and 11 and Voyagers 1 and 2—that are headed out of the solar system and that may soon encounter the termination shock. However, only the two Voyagers are still performing science operations. Courtesy of the Jet Propulsion Laboratory.

FIGURE 1.2 The Sun's corona imaged in the extreme ultraviolet (171 Å) by the TRACE telescope. The emissions are from Fe IX/X at a temperature of ~1,000,000 K. The magnetic field emerging from the photosphere structures the coronal plasma in an intricate and dynamic architecture of loops, arcades, and filaments. Courtesy of NASA and the Stanford-Lockheed Institute for Space Research.

As the solar wind flows away from the Sun and fills the heliosphere, it interacts in various and complex ways with the planets and other solar system bodies that it encounters. The nature of this interaction depends critically on whether the object has an internally generated magnetic field (Mercury, Earth, the giant outer planets) or not (Venus, Mars, comets, the Moon). For example, Mars has no strong global magnetic field, and the solar wind impinges directly on a significant fraction of its thin carbon dioxide atmosphere.[6] The erosion of the atmosphere resulting from this interaction may have played an important role over the last 3 billion or so years in the evolution of Mars's atmosphere and climate.[7] In contrast, the terrestrial atmosphere is protected from direct exposure to the solar wind by Earth's magnetic field, which forms a complex and dynamic structure—the magnetosphere—around which most of the solar wind is diverted (Figure 1.3). The solar wind's interaction with the magnetosphere, effected principally through a temporary merging of the interplanetary and terrestrial magnetic fields, stirs and energizes the magnetospheric plasmas and leads to periodic explosive releases of magnetic energy. In these events, known as magnetospheric substorms, powerful electrical currents flow between the magnetosphere and the ionosphere, injecting several billion watts of power into the upper atmosphere and producing often quite spectacular displays of the aurora borealis and australis—the Northern and Southern Lights (Figure 1.4). Unfortunately, at times of extreme magnetospheric disturbance, the beauty of the auroras can be accompanied by less benign phenomena—radiation belt enhancements, for example, and ionospheric disturbances—that can disrupt the communications, navigation, and power systems on which modern society so extensively depends.

Earth's aurora is ultimately powered by energy produced in thermonuclear reactions within the Sun's core and conveyed to Earth by the Sun's magnetized wind.[8] The last half century has seen remarkable advances in our knowledge and understanding of this stellar wind and its solar source, and of its interactions with Earth, other planets, and the ionized and neutral gases of the local interstellar medium. The beginnings of this understanding reach farther into the past, however, to the middle of the 19th century, when evidence for connections between solar activity and terrestrial phenomena began to accumulate rapidly and the work of James Clerk Maxwell unified the foundations for the theoretical description of the electromagnetic forces that govern the behavior of matter in the plasma state.

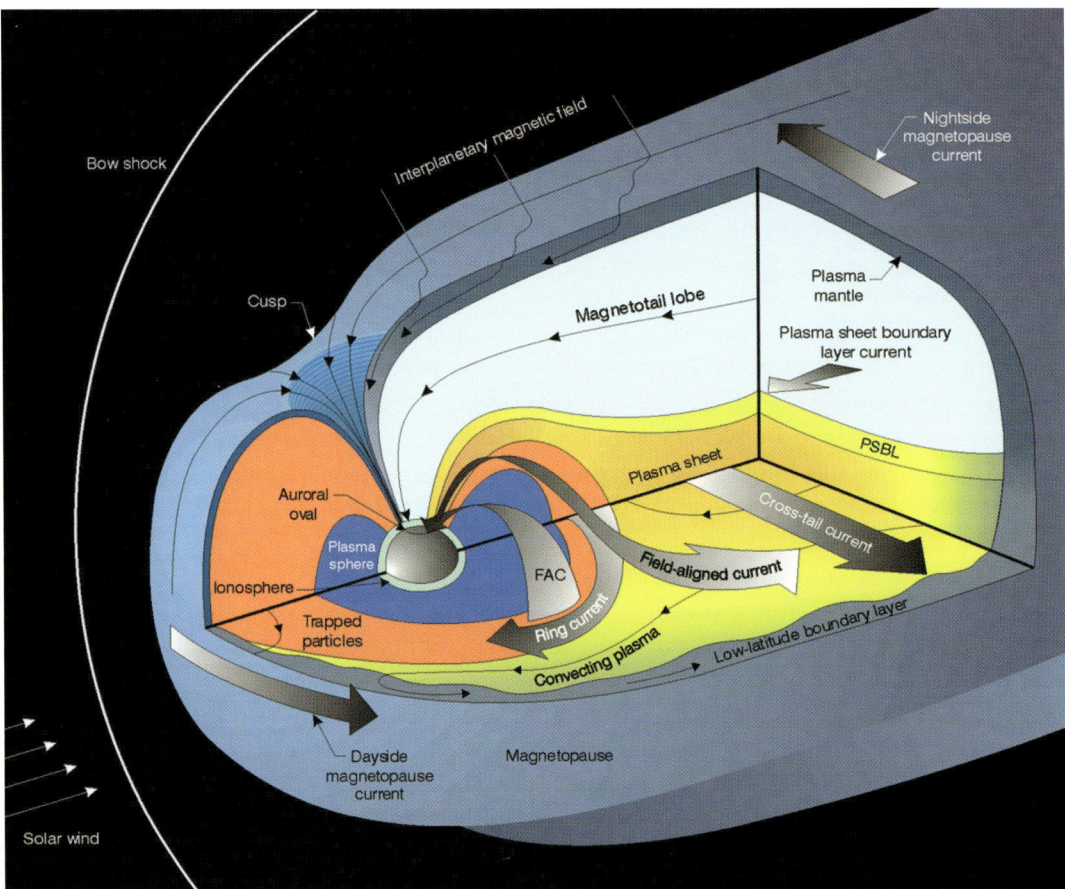

FIGURE 1.3 Artist's conception of Earth's magnetosphere, the volume of space around Earth dominated by the geomagnetic field and populated with plasmas of both ionospheric and solar wind origin. These plasmas are organized in distinct structures or regions characterized by different properties and separated by extremely thin boundary layers. A current sheet known as the magnetopause separates the magnetospheric plasma from the solar wind. In the upstream direction, this boundary is located at an average distance of 10 Earth radii (1 Earth radius = 6,378 kilometers) from Earth's center. Farther upstream, a standing shock wave (the bow shock) forms as the supersonic wind is slowed and heated by its encounter with the magnetosphere. On the nightside, the interaction with the solar wind stretches the terrestrial field into an elongated, tail-like structure that extends hundreds of Earth radii—well beyond the orbit of the Moon—in the antisunward direction. The ionized gases that populate the magnetosphere are remarkably dilute: The densest magnetospheric plasma is 10 million times less dense than the best laboratory vacuum! Nevertheless, the motions of these highly tenuous plasmas drive powerful electrical currents, and during disturbed periods, Earth's magnetosphere can dissipate well in excess of 100 billion watts of power—a power output comparable to that of all the electrical power plants operating in the United States.

MILESTONES AND SCIENCE CHALLENGES

FIGURE 1.4 Earth's aurora as seen in visible light from the ground (left) and in the far ultraviolet from the Earth-orbiting IMAGE spacecraft (right). Auroras are produced when energetic charged particles from the magnetosphere precipitate into Earth's atmosphere. The charged particles collide with the atoms and molecules of the upper atmosphere, exciting them and causing them to emit light at various wavelengths. The ground-based photograph is courtesy of Jan Curtis. The space-based image is courtesy of the Imager for Magnetopause-to-Aurora Global Exploration (IMAGE) far-ultraviolet imaging team and NASA.

MILESTONES: FROM STONEHENGE TO SOHO

The Sun and the heavens have long occupied a special place in human culture, as is amply documented by archaeological and other evidence of the astronomical interests of many different "pre-scientific" peoples, from the Neolithic builders of Stonehenge to the Mayans, whose astronomer-priests prepared solar-eclipse prediction tables of amazing accuracy. It was not until the 19th century, however, some 200 years after the invention of the telescope and the discovery of sunspots, that the systematic scientific study of the Sun began, with the discovery of the sunspot cycle and the application of the new science of spectroscopy to the analysis of solar composition. During the second half of the century, correlations reported between solar activity (as manifested in the changing sunspot number and in flares), disturbances in the Earth's magnetic field, and auroral activity clearly suggested the existence of a physical connection between the Sun's activity and terrestrial magnetic and upper atmospheric phenomena. The nature of this connection—one of the central themes of space physics—became the subject of intense study and controversy during the first half of the 20th century. By midcentury, the prevailing theory involved ionized "corpuscular streams" from the Sun that traveled at speeds of 1,000 to 1,600 kilometers per second and within which the geomagnetic field formed a cavity.[9] This picture was changed dramatically in the late 1950s when it was shown theoretically that the outer solar corona could not be static but must be continually expanding outward. The model of individual corpuscular streams was replaced by the modern concept of a continuous solar wind.[10]

The launch of the first Earth-orbiting satellites in the late 1950s and of the first interplanetary probes a few years later revolutionized the scientific community's ability to study the Sun-Earth connection, the interplanetary environment, and the space environments of the other planets. The event that signaled the opening of the new era in the exploration of space was the startling discovery, in 1958, of belts of trapped energetic charged particles (the Van Allen belts) circling Earth's middle and low latitudes at altitudes between 400 km and 60,000 km.[11] Within a short time of this discovery, the theoretically predicted solar wind with its embedded magnetic field had been observed and measured, and the cavity within it—now named the magnetosphere—was being surveyed. In situ measurements from a series of Earth-orbiting spacecraft, operating both inside and outside the magnetosphere, mapped this region's large-scale structure, establishing the existence of a boundary (the magnetopause) separating the magnetosphere from the solar wind and of a twin-lobed magnetic tail extending many hundreds of Earth radii in the antisunward direction. Theory had predicted that the supersonic nature of the solar wind would give rise to a collisionless bow shock upstream of the magnetosphere, and this prediction was verified by in situ satellite observations. These early space missions also confirmed the presence inside the magnetosphere of a ring current and of magnetic-field-aligned currents flowing between Earth's high-latitude upper atmosphere and the magnetosphere.[12]

Of fundamental importance for the field of solar-terrestrial research were the prediction[13] and discovery during the first decade of the space age of a link between geomagnetic activity and the orientation of the magnetic field embedded in the solar wind (the IMF). During the ensuing decades, space physicists made significant progress in understanding this link, which involves the merging of the interplanetary and terrestrial magnetic fields and the consequent transfer of energy, mass, and momentum from the solar wind into the magnetosphere, often resulting in major disturbances of Earth's space environment. Another important milestone in the study of the Sun-Earth connection has been the discovery of solar energetic particle events. Early direct balloon- and ground-based measurements had shown the occasional existence of energetic particles of solar origin in addition to the ever-present flux of galactic cosmic rays. In the late 1950s and early 1960s, measurements of cosmic radio noise absorption at high geomagnetic latitudes revealed that these events were more frequent than had been realized, and subsequent in situ measurements from balloons, rockets, and satellites have provided a wealth of information on the properties of these energetic particle fluxes, which form an important component of space weather and its human consequences.

The picture of Earth's space environment that has emerged from more than 45 years of in situ measurements and correlative ground-based observations, supported by increasingly sophisticated theoretical and modeling studies, is one of a highly complex, tightly coupled system comprising the ionosphere and thermosphere as well as the various magnetospheric plasma regions and boundaries. Thanks to a new generation of high-resolution instrumentation, space physicists are now beginning to discern the detailed workings of the nonlinear physics that plays such an important role in this coupled magnetosphere-ionosphere-atmosphere system. Complementing the ability to probe key plasma physical processes in situ are new imaging techniques, such as neutral atom imaging of the ring current and extreme-ultraviolet imaging of the plasmasphere, which allow researchers to view for the first time the global structure of the magnetosphere and its dynamical response to changing solar wind inputs (Figure 1.5). Of continuing importance in the study of the terrestrial space environment is the worldwide network of ground-based instruments—magnetometer chains, incoherent scatter radars, riometers, all-sky imagers—that provide invaluable contextual information to complement in situ spacecraft data.

The reconnaissance—and, in two cases, the exploration—of planetary environments has gone hand in hand with the intensive study of Earth's magnetosphere. Flyby missions have provided brief but invaluable snapshots of the space environments of all the planets except Pluto. The 1970s witnessed the first direct observation of Jupiter's magnetosphere, the existence of which had been revealed almost 20 years earlier by the detection of nonthermal radio emissions, and the discovery of the magnetospheres of Mercury and Saturn. The following decade saw the discovery of the magnetospheres of Uranus and Neptune, which were found to have complex geometries and plasma dynamics owing to large tilt angles between the planetary magnetic and rotational axes. Increasing knowledge of the prop-

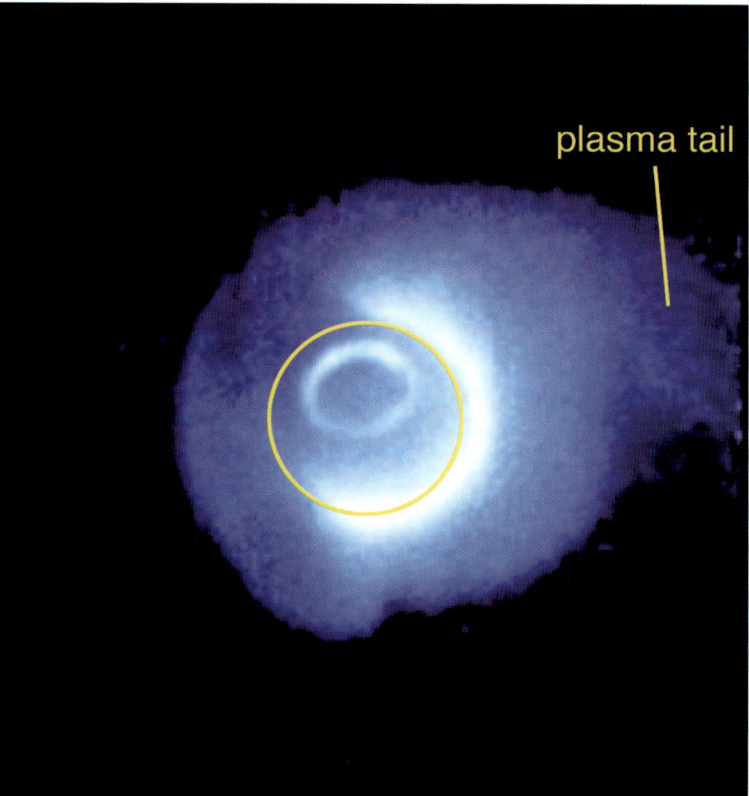

FIGURE 1.5 Image of Earth's plasmasphere, created by a new imaging technique that detects extreme ultraviolet (EUV) sunlight (30.4 nm) resonantly scattered from singly charged helium ions. The yellow circle indicates the position of Earth; the sunward direction is to the lower right. The plasmasphere, which appears in this image as a pale blue cloud surrounding Earth, is a region of cold (1 eV), dense (10^3 cm^{-3}) plasma that corotates with Earth. Visible in the image is a faint tail of plasma that is being eroded from the plasmasphere and transported to the dayside magnetopause, where it is lost into the solar wind. An active northern aurora can also be seen. Such techniques as EUV and energetic neutral atom imaging make it possible to visualize the global structure and dynamics of magnetospheric regions that previously could be studied only through in situ measurement. Courtesy of the IMAGE EUV imaging team and NASA.

erties of other magnetospheres provides a basis for comparative magnetospheric studies in which theoretical concepts developed in studies of Earth's magnetosphere can be tested, refined, and expanded.

In addition to the reconnaissance of planetary magnetospheres during flyby encounters, space physics measurements have been made on spacecraft orbiting Venus, Jupiter, and, most recently, Mars. For example, extensive explorations of two quite different planetary environments have been carried out by spacecraft in orbit around Venus and Jupiter. The 14-year Pioneer Venus Orbiter (PVO) mission gave researchers a detailed look at the interaction between an unmagnetized planet and the solar wind. PVO furnished a wealth of information about the complex processes that characterize such interactions—processes such as the mass loading of the solar wind through the entrainment of ionized atmospheric material and the formation of induced magnetotails. More recently, the survey of the Jupiter system by the Galileo spacecraft has provided important new data about the workings of a magnetosphere that is fundamentally different from Earth's in terms of both its primary plasma source (volcanic emissions from the moon Io) and its energy source (planetary rotation as opposed to the solar wind, as at Earth). Moreover, results from the Galileo mission have added a new "species" to the taxonomy of solar system and astrophysical magnetospheres through the discovery of Ganymede's magnetic field, which forms a minimagnetosphere within the massive magnetosphere of Jupiter.

The second half of the 20th century, and in particular its last quarter, witnessed important advances in the study of the Sun and the heliosphere. Two major developments in solar physics during the 1970s were the detailed characterization of coronal holes and the discovery of coronal mass ejections. Coronal holes, low-density regions of open magnetic field, were shown to be the source of the high-speed solar wind, which dominates the heliosphere near and at solar minimum. As noted above, coronal mass ejections are powerful eruptions of coronal plasma and magnetic fields. They occur most frequently at solar maximum and are responsible for nonrecurrent geomagnetic storms.

Much of the progress in the study of the solar corona has been made possible by the development of the capability to observe the Sun at ultraviolet and x-ray as well as visible wavelengths, first from sounding rockets and later from space-based observatories. Three such observatories launched in the 1990s—Yohkoh, the Solar and Heliospheric Observatory (SOHO), and TRACE—have recorded the changes in the Sun from the declining phase of solar cycle 22 through the activity peak of cycle 23 and have uncovered important clues about the vital role of magnetic reconnection in the rapid release of energy on the Sun, the complex fine structure and dynamic nature of the corona, and the nature of the coronal heating process.

In addition to new insights into the physics of the corona, the last decade or so has also seen notable advances in our knowledge of the Sun's interior. These advances are attributable to the development of helioseismology, which has given solar physicists a powerful tool with which to probe the internal structure and dynamics of the Sun (see sidebar, "Helioseismology"). A major achievement in helioseismology has been the observational confirmation of theoretical models of the structure of the Sun's interior. Helioseismic measurements have advanced our understanding of the motions of the Sun's interior as well as of its structure, revealing, for example, large-scale meridional flows in the outer layer of the convection zone, the signatures of newly emerging active regions beneath the solar surface, and the existence of strong velocity shears at the base and top of the convection zone. It is in these last-mentioned shear layers that the Sun's large-scale magnetic field (lower layer) and small-scale magnetic field (upper layer) are likely to originate.

Since the first definitive solar wind observations by Mariner 2 on its way to Venus in 1962, the heliosphere has been studied in situ by a series of spacecraft and remotely by interplanetary radio scintillation and other remote-sensing techniques. The average properties and structure of the solar wind and interplanetary magnetic field in the ecliptic plane and at 1 AU were well established by the early 1970s and were shown to be highly variable. Much of the solar wind's variability was found to be either recurrent at the 27-day solar rotation period or episodic in association with disturbances on the Sun. In situ study of the heliosphere beyond 1 AU became possible with the launch in the 1970s of the Pioneer and Voyager outer planetary probes. Research topics in heliospheric physics on which significant advances were made during the 1970s and 1980s included the propagation of transient disturbances and turbulence through the heliosphere, the corotating interaction regions (CIRs) formed by alternating high- and low-speed solar wind streams during the declining phase of the solar cycle and at solar minimum, the acceleration of energetic particles at heliospheric shocks, and the distribution and modulation of cosmic rays in the outer heliosphere (see sidebar, "Galactic and Anomalous Cosmic Rays").

A landmark development in heliospheric studies occurred in 1990 with the launch of the joint European Space Agency (ESA)/NASA Ulysses mission. Ulysses' polar orbit has given researchers their first in situ look at the high-latitude heliosphere, and its long operational lifetime has permitted detailed study of the solar wind from the declining phase of the solar cycle

HELIOSEISMOLOGY

Helioseismology is the study of the structure and dynamics of the Sun's interior through the observation of oscillations on the visible solar surface (the photosphere). Such oscillations are produced by acoustic (p-mode) waves generated at the top of the solar convection zone. By analyzing the spectrum of these oscillations, helioseismologists are able to measure the sound speed, density, and bulk velocity in the interior of the Sun as a function of depth and can determine the departures of these quantities from spherical symmetry. The sound speed is now known to about one part in 10^4 through most of the solar interior.

A major advance in our ability to perform helioseismic studies occurred in the mid-1990s, with the organization of global networks of dedicated observing sites, preeminent among which are the six sites of the Global Oscillations Network Group (GONG) of the National Optical Astronomy Observatories and the deployment of spaceborne helioseismology instruments on the Solar and Heliospheric Observatory (SOHO), a joint ESA-NASA mission. GONG and SOHO allow helioseis-mologists to observe the Sun continuously over long periods of time, which is necessary to resolve the global normal-mode spectrum of the oscillations. SOHO observations, moreover, are unaffected by atmospheric distortion and thus can achieve the spatial resolution required to measure the higher-degree modes, which provide information about the detailed dynamics of the convection zone.

In addition to global helioseismology, which focuses on the study of large-scale structures, "local" helioseismological techniques are now being developed to probe structures and flows below the surface of smaller-scale phenomena on the Sun. One such technique—"heliotomography"—has been successfully employed to provide the first images of the subsurface structure of a sunspot (see figure).

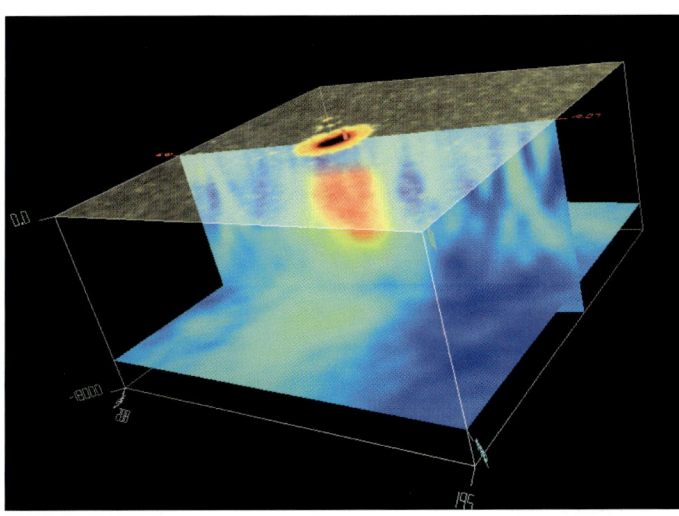

The subsurface structure of a sunspot as derived from SOHO/MDI data through the use of heliotomographic inversion techniques. Courtesy of the SOHO/MDI consortium.

GALACTIC AND ANOMALOUS COSMIC RAYS

Cosmic rays are ubiquitous, existing throughout the heliosphere and the universe. Because of their mobility, speed, and responsiveness to electromagnetic fields, cosmic rays can serve as probes of otherwise inaccessible regions of the heliosphere and the Galaxy. Observations of cosmic rays and sophisticated theory have yielded many major discoveries and have significantly advanced our understanding of the Sun, heliosphere, and the Galaxy.

Galactic cosmic rays (GCRs) are energetic (from hundreds of MeV to GeV) charged particles entering from the Galaxy. The majority of GCRs are accelerated at shocks produced by supernova explosions. As they propagate through the heliosphere, their intensity and properties are modulated by the structure of the solar wind. This modulation is seen, for example, in the anticorrelation between solar activity and GCR intensity (see figure on the facing page).

Anomalous cosmic rays (ACRs) were first observed in the early 1970s as a peculiar, or "anomalous," distortion of the cosmic-ray composition. Most ACRs begin as galactic neutral atoms that enter the heliosphere and are subsequently ionized. Upon ionization, they are picked up by the solar wind's magnetic field and carried outward, toward the termination shock. Some are accelerated to higher energies by interplanetary shocks and may experience further acceleration—to energies in excess of 1 GeV—at the termination shock. Because of their mobility, the ACRs can then move throughout the heliosphere, to be observed as a significant distortion of the observed galactic cosmic-ray spectrum.

MILESTONES AND SCIENCE CHALLENGES 39

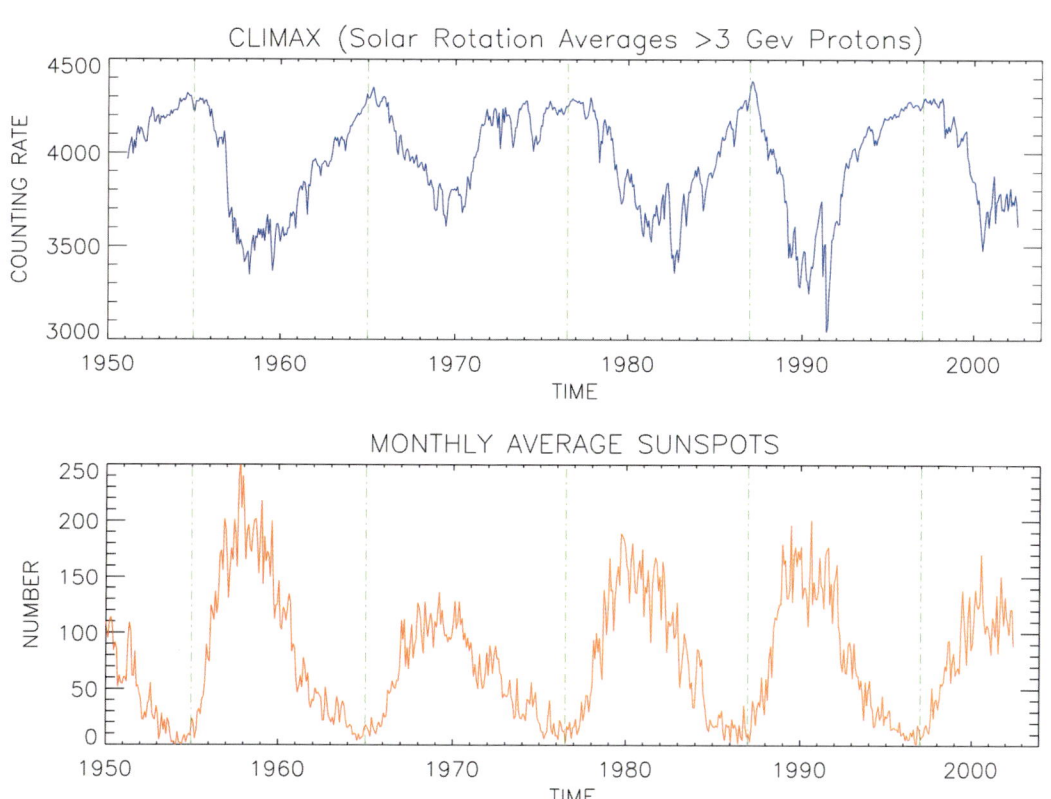

A comparison of galactic cosmic ray intensity, as determined from counting rates at the University of Chicago's neutron monitor in Climax, Colorado (top panel) and monthly average sunspot numbers (bottom panel). The data, which cover a period of more than five decades, from the declining phase of solar cycle 18 through the peak of solar cycle 23, clearly illustrate the anticorrelation between galactic cosmic ray intensity and solar activity. Courtesy of J.R. Jokipii (University of Arizona), the University of Chicago, the National Science Foundation, and the World Data Center for the Sunspot Index (Brussels, Belgium).

through solar minimum to solar maximum. Ulysses' observations have shown, for example, that the heliosphere is dominated by the fast solar wind from high latitudes at solar minimum but by a slow and variable wind from all latitudes near and at solar maximum. Ulysses has revealed the three-dimensional structure of CIRs and observed the propagation of solar energetic particles to high heliographic latitudes. Ulysses has also contributed important new information about interstellar material in the heliosphere, making the first measurements of several species of interstellar pickup ions and observing their acceleration to high energies at CIR-associated shocks.

The preceding review of milestones in solar and space physics has of necessity been highly selective and has focused on developments resulting from improved observational capabilities. It must be emphasized, however, that a number of major milestones in the history of solar and space physics have been in the area of theory. Indeed, theory in several cases anticipated what observation later confirmed. Three important examples have already been cited: the theory of the solar wind as a hydrodynamic phenomenon originating in coronal expansion, the prediction of the formation of a collisionless bow shock upstream of Earth's magnetosphere, and the model of the merging of the interplanetary and terrestrial magnetic fields in controlling plasma transport and acceleration in Earth's magnetosphere. In heliospheric physics, theory also predicted the existence of the spiral interplanetary magnetic field; the interplanetary magnetic field's role in modulating cosmic ray intensity at Earth; the overall large-scale structure of the heliosphere, including the termination shock (not yet directly observed); the corotating interaction regions between fast and slow solar wind streams; the development of forward and reverse shocks bounding those interaction regions at large heliocentric distances; and transient shock wave disturbances driven by the episodic ejection of high-speed plasma from the Sun. Among the successful theoretical predictions in magnetospheric and ionospheric physics are the existence of the magnetopause, the existence of the extended magnetospheric tail, the presence of Birkeland or magnetic-field-aligned currents coupling the magnetosphere with the ionosphere, and the existence of the polar wind. Other milestones in theoretical solar and space physics include the advanced theory of cosmic-ray transport and the alternating flat and sharp top cosmic-ray maxima, plasma convection within planetary magnetospheres, shock acceleration of energetic particles, magnetic reconnection, and magnetohydrodynamic turbulence. Although first formulated to account for phenomena within the immediate domain of solar and space physics, many of these theories address physical processes that are fundamental and ubiquitous and provide a foundation for understanding other astrophysical systems.

SCIENCE CHALLENGES

The accomplishments of the past decades have answered important questions about the physics of the Sun, the interplanetary medium, and the space environments of Earth and other solar system bodies. However, they have also highlighted other questions, some of which are long-standing and fundamental. The committee has organized these questions in terms of five "challenges," which are presented below, along with some of the representative questions that will be the focus of scientific investigations during the coming decade and beyond. Answers to these remaining questions have so far eluded our grasp owing to the lack of observations in critical regions of the solar system, limitations on the capability of our observational techniques and strategies to resolve critical processes, constraints on computational resources and techniques, and various infrastructural and programmatic issues. Fortunately, however, none of these impediments is insurmountable, and the solar and space physics communities look forward to the implementation, during the first decade and a half of the new millennium, of missions, facilities, and programs that will equip them to respond to the five scientific challenges set forth below and finally to answer some of the outstanding questions about the Sun and the objects immersed in and interacting with its atmosphere.

The Sun's Dynamic Interior and Corona

Challenge 1: Understanding the structure and dynamics of the Sun's interior, the generation of solar magnetic fields, the origin of the solar cycle, the causes of solar activity, and the structure and dynamics of the corona. Helioseismic sounding of the solar interior has led to significant advances in stellar evolution theory. These advances have come through successively more precise seismic soundings and better determinations of the properties of the plasma inside the Sun. Solar physicists can now explore the interior immediately beneath sunspots and can expect over the next decade to accurately sound the base of the solar convection zone, revealing the detailed nature of turbulent convection. This knowledge ahould advance understanding of the nature of stars with convective zones.

The dynamo problem is a fundamental and unsolved problem in physics. Without a full description of the solar dynamo, our description of the

Sun and various heliospheric processes powered by the solar magnetic field remains incomplete. An understanding of magnetic field generation is also the key to understanding magnetic activity in whole classes of solarlike stars. The concentrated fibril bundles of emerging subsurface solar magnetic flux are mysterious and far from fully explored. Periodic reversals of magnetic field, variability of the level of activity, and the nature and origin of coronal holes, all features of the solar cycle, are fundamental drivers of the state of the entire heliosphere.

In recent years, the lower corona has been observed at spatial and time resolutions that have revealed details of phenomena not yet explained in terms of basic physics. It is still not understood how extremely high temperatures, in excess of several million degrees, are maintained above the cooler solar surface. Although the energy that powers this heating must come ultimately from the convection zone and photospheric activity, we do not understand how it is transported through the chromosphere and into the corona and how it is converted to heat beginning within about a tenth of a solar radius above the photosphere. This intellectual deficit directly translates into an incomplete understanding of the origins of the solar wind and of the baseline conditions for establishing the entire extended solar atmosphere. Although the Ulysses mission established a firm distinction between two different states of the solar wind (fast, hot, rarefied and slow, cooler, denser), the origins of the two states remain important questions, and it is not known if the slow wind is quasi-stationary or results from transient processes. Determining the composition and evolution of solar wind plasma structures remains a major challenge.

One of the most significant developments of the past decade has been the recognition of the importance of CMEs in driving geoeffective heliospheric disturbances. Many questions remain, however, about the initiation and evolution of CMEs—about the role of magnetic reconnection, for example, and about the origin of the magnetic flux rope structure observed in a number of CMEs. Similarly, the past decade has seen notable advances in our understanding of particle acceleration at the Sun and in the heliosphere; of particular importance is the recognition of the role of both flares and CME-driven shocks in the acceleration of solar energetic particles (SEPs). However, much remains to be learned about the spatial and temporal evolution of the SEP sources and about the basic SEP acceleration and transport processes.

Key questions include:

- How and where does the solar dynamo operate, and in what ways do the fields created by the dynamo move up through the visible surface? How do the magnetic structures in the corona follow?
- What physical processes are responsible for heating the active x-ray corona? What heats the coronal holes?
- What are the basic physical mechanisms for, and characteristics of, the acceleration of the fast and slow solar winds? What controls the development and evolution of the solar wind in the inner heliosphere?
- What is the physics of explosive energy release in the solar atmosphere? How are CMEs initiated? How and where are particles accelerated at the Sun?

The Heliosphere and Its Components

Challenge 2: Understanding heliospheric structure, the distribution of magnetic fields and matter throughout the solar system, and the interaction of the solar atmosphere with the local interstellar medium. The heliosphere is not static. Complex and time-dependent phenomena such as stream interfaces, multiple current sheets, and CMEs are pervasive. Flux tubes are set into motion by coronal and photospheric processes that are not well understood. Interplanetary spacecraft instruments have repeatedly identified turbulence in the solar wind, but as yet a complete picture of the origin, evolution, and distribution of turbulent fluctuations is lacking. Although the interplanetary magnetic field and the radial distribution of plasma are understood at a coarse level, there are important unanswered questions. The structure of the magnetic field at high latitudes is unclear. The winding of the spiral magnetic field at large distances is not consistent with models. The distributions of dust, anomalous cosmic rays, and pickup ions all promise to reveal new physical insights about the overall structure of the heliosphere. The propagation of solar energetic particles and galactic cosmic rays depends on effects at the boundaries and also on small-scale turbulence. Thus the problems of cosmic ray modulation and energetic particle transport are particularly challenging, as they require a systemic understanding of heliospheric properties.

Although direct exploration of interstellar space remains an important future goal, at present the heliopause bounds the domain of space physics and presents a more immediate goal for exploration. The complex solar atmosphere engenders an equally complex interaction with the interstellar medium. Although not yet directly explored, this distant boundary produces remote effects that we have begun to observe and study. For example, interstellar neutral atoms do not respond to the interplanetary magnetic field and thus are able to penetrate into the heliosphere, producing observable effects. These neutrals become ionized by charge exchange or ultraviolet radiation and become pickup ions that respond to and influence the solar wind and its magnetic field. The pickup ions are accelerated by interplanetary shocks and again to even higher energies at the termination shock, becoming the anomalous cosmic rays observed at Earth. A fascinating plasma physics process in its own right, this series of events and the feedback associated with it are facets of the spatially distributed interaction between the interplanetary and interstellar media. The transition at the heliopause between magnetic fields of solar origin and those of interstellar origin is another such boundary effect. As we begin to understand these interactions better in the coming decade, we will also begin to learn more about the next frontier of exploration, the local interstellar medium and the nearby parts of the galaxy that are not in direct plasma contact with our Sun.

Key questions include:

- How do coronal structures evolve into solar wind structures of varying speed, density, kinetic temperature, composition, and magnetic field strength? How do CME-driven disturbances evolve in space and time as they propagate through the heliosphere?
- What is the structure of the interplanetary magnetic field at very large distances from the Sun and as a function of the solar cycle? How are plasma, neutrals, heavy ions, turbulent fluctuations, solar energetic particles, and galactic cosmic rays distributed throughout the entire heliospheric volume?
- How do the solar wind plasma and magnetic field interact with the electromagnetic field, plasma, and neutrals in the nearby region of the galaxy? How and where is the boundary of the heliosphere established? How does it move in time, and how do such changes affect our space environment? What is the nature of the local interstellar medium?

Space Environments of Earth and Other Solar System Bodies

Challenge 3: Understanding the space environments of Earth and other solar system bodies and their dynamical response to external and internal influences. Of the magnetized solar system bodies, it is Earth's space environment and its interaction with the solar wind that have been most extensively studied. As discussed above, more than 40 years of space- and ground-based observations have provided a good picture of the overall structure of the magnetosphere and its constituent plasma populations, of the solar wind's role in driving magnetospheric dynamics, and of the electrodynamic coupling of the magnetosphere to the ionosphere. Nonetheless, important gaps in scientific understanding remain—concerning, for example, the configuration and dynamics of the magnetosphere under extreme solar wind conditions (i.e., during strong geomagnetic storms); many aspects of magnetic reconnection at the dayside magnetopause and in the magnetotail; particle energization in the inner magnetosphere; the complex structure of the magnetotail as it absorbs and releases energy extracted from the solar wind; and the temporal and spatial scales of the electrodynamical processes by which some of this energy, along with momentum, is transferred to and redistributed within the ionosphere-thermosphere system. Much remains to be learned about the dynamics and energetics of Earth's middle atmosphere and its coupling to the ionosphere-thermosphere system. Poorly understood, too, are possible influences of the space environment on Earth's weather and climate.

Jupiter's magnetosphere is the most thoroughly studied magnetosphere after Earth's. Six flyby missions and Galileo's successful 8-year tour, complemented by high-resolution auroral imaging from both ground- and space-based observatories, have yielded a wealth of information about the rotationally powered Jovian magnetosphere. However, Jupiter's high-latitude magnetosphere, a region of crucial importance for understanding the transfer of planetary rotational energy to the magnetosphere and the transfer of magnetospheric energy back into the upper atmosphere, remains unexplored.

The next decade will see extensive exploration of the magnetosphere of another giant planet, Saturn. Following its arrival in the Saturn system in 2004, the Cassini spacecraft will gather data on Saturn's magnetospheric plasma sources (rings, icy satellites) and make observations needed to assess the relative contributions of solar wind energy and planetary rotational energy to dynamics.

Of particular interest in the Saturn system is the interaction between the moon Titan, with its massive nitrogen-methane atmosphere, and the plasma contained in Saturn's magnetosphere. Observations made during Cassini's many flybys of Titan will deepen our knowledge of the processes involved in the interaction between an unmagnetized or weakly magnetized body and an externally flowing magnetized plasma and the effects of this interaction on the structure and dynamics of the body's atmosphere. Such processes, already extensively studied at Venus by the Pioneer Venus Orbiter, will also be investigated during the next decade at Mars by the Japanese Nozomi mission and the European Mars Express mission. Owing to the presence of localized crustal remanent magnetic fields, Mars's interaction with the solar wind is expected to be more complicated than that of Venus with the solar wind and that of Titan with Saturn's magnetospheric plasma.

The smallest of the planetary magnetospheres is Mercury's, for which only limited data, acquired by Mariner 10 during two passes in the mid-1970s, are available. These data indicate that Mercury's magnetosphere, like Earth's, is energized by the interaction with the solar wind and suggest that intense, substorm-like particle acceleration events occur in its magneto-tail. It is not understood, however, how Mercury's magnetosphere, in the absence of a conducting ionosphere, dissipates the energy that it extracts from the solar wind. This question will be addressed by two Mercury orbiter missions, NASA's MESSENGER mission and the ESA's BepiColombo mission, both scheduled for launch within the next decade.

Key questions include:

• Does the state of Earth's magnetosphere under extreme solar wind conditions differ qualitatively as well as quantitatively from its state under more moderate conditions? How and where does magnetic reconnection occur on Earth's dayside magnetopause? What processes are responsible for the rapid acceleration of charged particles to hundreds of keV at the onset of magnetospheric substorms and to MeV energies during geomagnetic storms? Do the active auroral displays during substorms arise from instabilities in the ionosphere, or do they simply mirror plasma motions in the outer magnetosphere? To what extent are the ionized and neutral gases of Earth's upper atmosphere affected by mechanical and electrodynamic inputs from the lower atmosphere?

• How does angular momentum conservation alter the dynamics of reconnection in Jupiter's rapidly rotating magnetosphere? How is Jupiter's high-latitude ionosphere coupled to the plasma populations of the magnetodisk? How do field-aligned currents arising from the interaction of moons with Jupiter's magnetospheric plasma produce the emissions observed in the high-latitude ionosphere? What process is responsible for the pulsating x-ray aurora at Jupiter?

- What is the effect of localized crustal magnetic fields on the solar wind interaction with Mars? How are outflowing ions on the nightside of Mars accelerated to keV energies?
- How does the absence of a significant ionosphere affect the interaction of Mercury's ionosphere with the solar wind? How do substorms at Mercury differ from terrestrial substorms?

Fundamental Space Plasma Physics

Challenge 4: Understanding the basic physical principles manifest in processes observed in solar and space plasmas. The ultimate goal of solar and space physics research is not merely to produce a detailed phenomenological description of its various objects of study but also to understand the fundamental physical processes that operate in them. As these processes largely involve matter in the plasma state, solar physics and space physics can both be considered branches of plasma electrodynamics, to which field they have contributed significantly over the years. The heliosphere is a natural laboratory for the study of plasma physics, and the next decade of research can be expected to lead to advances in our understanding of such fundamental plasma physical processes as magnetic reconnection; turbulence; charged particle acceleration and scattering; generation, transport, and damping of plasma waves; and magnetic dynamo action. The study of these processes in naturally occurring solar system plasmas will be complemented by their investigation with increasingly sophisticated computer models and in laboratory plasma experiments (see Chapter 4).

Key questions include:

- What controls the rate of collisionless magnetic reconnection? How is reconnection initiated? Does it occur spontaneously or episodically or must it be driven by external triggers? What are the roles of non-magnetohydrodynamic[14] processes (kinetic Alfvén waves, whistler-mode waves, electron inertial effects) in magnetic reconnection?
- What is the nature of turbulence in nonuniform plasmas? How does microturbulence couple to magnetohydrodynamic or fluid turbulence? How does turbulence propagate across plasma boundaries?
- What are the processes by which particles are accelerated to very high energies in the heliosphere, and what governs their transport?
- What are the conditions under which the electrodynamic interaction of a conducting body with an ambient magnetized plasma generates waves that can affect the particle population?

Space Weather

Challenge 5: Developing near-real-time predictive capability for understanding and quantifying the impact on human activities of dynamical processes at the Sun, in the interplanetary medium, and in Earth's magnetosphere and ionosphere. Space weather describes the conditions in space that affect Earth and its technological systems. Space weather is a consequence of the behavior of the Sun, the nature of Earth's magnetic field and atmosphere, and our location in the solar system. Through various complex couplings, the Sun, the solar wind, and the magnetosphere, ionosphere, and thermosphere can influence the performance and reliability of spaceborne and ground-based technological systems. Solar energetic particle events and geomagnetic storms are natural hazards, like hurricanes and tsunamis. Solar energetic particle events can disrupt spacecraft operation and present a radiation hazard to astronauts and to the crews and passengers of aircraft flying at high latitudes as well. Severe geomagnetic storms can interfere with communications and navigation systems, disturb spacecraft orbits because of increased drag, and cause electric utility blackouts over wide areas.

Both our understanding of the basic physics of space weather and our appreciation of its importance for human activity have increased considerably during the past decade. Much remains to be learned, however, about processes—such as radiation belt enhancements—that affect the environment in which many satellites operate; about the variations in the properties of the ionosphere-thermosphere system that can adversely affect Global Positioning System navigation systems and high-frequency radio wave propagation; and, finally, about the solar drivers of space weather. During the coming decade these problems will be the focus both of pure space physics research and of targeted basic research activities such as those envisioned in NASA's Living With a Star initiative, the National Science Foundation's National Space Weather Program, and the Department of Defense's *Space Weather Architecture Study*.[15] An important aspect of space weather-related research is the development of specification and predictive models that can be used for system design, space operations, and both now-casting and forecasting. Such models will be powerful and indispensable tools for the mitigation of the harmful effects of space weather.

Key questions include:

• What measurements need to be made to quantify the effects of space weather? Can a sustainable observing program provide the input needed

for useful predictions? Can the physics of space weather phenomena be understood well enough to become predictive?
- What requirements does the need to predict space weather place on specification models and on physics-based data assimilation models?
- What risks does space weather present for human spaceflight outside the protective shield of Earth's magnetosphere?

THE ASTROPHYSICAL CONTEXT

The above challenges outline, in broad brush strokes, major themes that characterize research objectives in solar and space physics. The committee recognizes that these disciplines ultimately belong to the broader intellectual enterprise of seeking to understand the universe at large and to comprehend our place in it. An additional far-reaching goal thus defines the larger context of all solar and space physics research:

Understanding the Sun, heliosphere, and planetary magnetospheres and ionospheres as astrophysical objects and in an astrophysical context. It is inevitable that we have come to a more specific and detailed understanding of our own star than of other stars. We also observe space plasma physics processes in the solar system much more closely than in more distant astrophysical objects. The detailed observations and models developed in space physics can therefore help to develop physically motivated explanations of less-well-constrained astrophysical phenomena. The cross-disciplinary influences flow in both directions, of course. That is, although we may acquire detailed knowledge about the Sun and its environment, it is only one star. If we believe that we have achieved an accurate understanding of a particular solar phenomenon, such as coronal heating, then we should be able to compare our model results with observations of Sun-like stars. Similarly, the physics of magnetospheres should have implications for astrophysical phenomena such as pulsars and jets (and, in the case of the jovian magnetosphere, for our understanding of the transfer of angular momentum by hydromagnetic processes in a rotating system). Moreover, many theoretical models would provide a way to scale physical predictions to stars that are not similar to the Sun. In this way, astrophysical observations can be used to test solar and space physics theories, and vice versa (see Chapter 4).

UNDERSTANDING COMPLEX, COUPLED SYSTEMS

Solar system plasmas are complex systems. Complexity arises from nonlinear couplings, both within a single system and between two or more different systems. Both types of coupling occur in solar and space plasmas. Beyond 10 to 15 AU, for example, the dominant constituent of the heliosphere by mass is neutral interstellar hydrogen. Charge exchange reactions couple this neutral hydrogen population and the solar wind, yielding a highly nonequilibrated, nonlinear system in which the characteristics of both populations are strongly modified. In addition to couplings between multiple constituents, solar system plasmas are characterized by couplings across a multiplicity of spatial and temporal scales; the nonlinear, dynamical, self-consistent feedback and coupling of all scales determine the evolution of the systems through the creation of large- and small-scale structures.[16] Examples of such cross-scale coupling are reconnection and turbulence, which involve the nonlinear interaction of large-scale, slow magnetohydrodynamic behavior and small-scale, fast kinetic processes. Finally, distinct plasma regions and regimes are coupled across boundaries in a highly nonlinear, dynamical fashion. Such cross-system coupling is exemplified by the coupling that occurs between the solar wind and Earth's magnetosphere as a result of the merging of interplanetary and geomagnetic field lines and by the electromagnetic coupling of the magnetosphere and the ionosphere.

The complex, nonlinear, coupled character of solar system plasmas presents significant challenges to both our observational capabilities and our theoretical understanding. The research initiatives recommended by the committee and presented in the following chapter will enable solar and space physicists to address those challenges and thereby to achieve new and deeper understanding of solar system plasmas and of the fundamental physical processes that govern them.

NOTES

1. Conversely, research and technology that are explicitly directed toward practical ends can make substantial contributions to "pure" scientific inquiry and the acquisition of fundamental knowledge. A classic example from the early days of the space age is the important role played by the Vela satellites in the exploration of the magnetosphere and the nearby interplanetary medium. Launched during the 1960s and in early 1970, the Velas were part of a joint program of the Department of Defense and the Atomic Energy Commission to monitor nuclear tests from space. (Besides their role in magnetospheric research, the Velas also made a major contribution to astrophysics, through their discovery of gamma ray bursts.) Operational satellites—those of the Defense Meteorological Satellite Program, the geosynchronous

spacecraft instrumented by the Los Alamos National Laboratory, and the meteorological satellites of the National Oceanic and Atmospheric Administration—continue to provide space physicists doing basic research with invaluable data about the terrestrial space environment. A recent example of the adaptation of an applied, real-world technology for the purpose of fundamental research is the use of data from the magnetometers on the Iridium communications satellites to prepare global maps of Earth's field-aligned current systems.

2. *SolarMax*, a 40-minute documentary in giant screen format, was written, directed, and produced by John Weiley. Chicago's Museum of Science and Industry is the film's executive producer and international distributor. NASA, the NSF, and the European Space Agency all assisted in the production of *SolarMax*.

3. The interface between the solar wind and the local interstellar medium is complex. It consists of three main structures: the heliopause, which separates the solar wind plasma and the interstellar plasma; the termination shock, a shock wave inside the heliopause where the solar wind speed changes from supersonic to subsonic; and the heliosheath, a region of shocked solar wind between the termination shock and the heliopause. The distances to the termination shock and the heliopause are not known and are most certainly variable, as described in the text. However, if estimates placing the shock between 85 and 100 AU from the Sun are correct, Voyager 1, traveling at a speed of 3.6 AU per year, may soon encounter it. (An astronomical unit is the mean distance between the Sun and Earth, roughly 150 million kilometers.) Depending on the distance to the heliopause, the spacecraft may also reach the boundary of the heliosphere—and perhaps even cross into the interstellar medium—before it no longer has sufficient power to operate its instruments (around 2020). Voyager 2 is currently at 67 AU and traveling at 3.3 AU per year; it will thus encounter the shock later than Voyager 1. Like Voyager 1, Voyager 2 will have enough power to operate its instruments until 2020.

4. Zank, G.P., and P.C. Frisch, Consequences of a change in the galactic environment of the Sun, *Astrophysical Journal* 518, 965-973, 1999.

5. See the discussion of solar energetic particle events in the National Research Council's *Radiation and the International Space Station: Recommendations to Reduce Risk*, National Academy Press, Washington D.C., 2000.

6. Partial protection of the martian atmosphere from the solar wind is provided by the strong, localized remnant crustal magnetic fields recently discovered by the Mars Global Surveyor magnetic field experiment. Because of the presence of these fields, the solar-wind/atmosphere interaction is considerably more complex at Mars than at Venus.

7. Solar-wind-driven atmospheric erosion has certainly influenced the evolution of Venus's atmosphere as well. However, the few studies of the implications of this loss process for the early history of the two planets have focused on Mars rather than Venus (e.g., Luhmann, J.G., R.E. Johnson, and M.H.G. Zhang, Evolutionary impact of sputtering on the Martian atmosphere by O+ pickup ions, *Geophysical Research Letters* 19, 2151, 1992, and Kass, D.M., and Y.L Yung, Loss of atmosphere from Mars due to solar wind-induced sputtering, *Science* 268, 697, 1995). It should be emphasized that such studies are subject to a number of important uncertainties—for example, those surrounding the level of the solar EUV flux, the planetary dynamo and global magnetic field at an earlier stage of Mars's history, and the effect on the solar-wind/atmosphere interaction of the remnant crustal fields.

8. Not all planetary magnetospheres are powered by the solar wind interaction. Jupiter's giant magnetosphere (the largest object in the solar system) draws its power primarily from the rotational energy of the planet. Saturn's magnetosphere, too, is thought to be rotationally driven. The relative roles of planetary rotation and the solar wind interaction in the dynamics of the Uranian and Neptunian magnetospheres are not known.

9. Chapman, S., and J. Bartels, *Geomagnetism*, Oxford University Press, 1940.

10. Parker, E.N., *Interplanetary Dynamical Processes*, Interscience Publishers, New York, N.Y., 1963.

11. In 1989, J.A. Van Allen (University of Iowa) was awarded the Crafoord Prize by the Royal Swedish Academy of Sciences for his discovery of the radiation belts. Van Allen describes the discovery of the radiation belts and the early days of magnetospheric research in *Origins of Magnetospheric Research*, Smithsonian Institution Press, Washington, D.C., 1983.

12. Useful surveys of the development of magnetospheric physics before and after the International Geophysical Year are given by D.P. Stern in his articles "A brief history of magnetospheric physics before the spaceflight era," *Reviews of Geophysics* 27, 103-114, 1989, and "A brief history of magnetospheric physics during the space age," *Reviews of Geophysics* 34, 1-31, 1991.

13. Dungey, J.W., Interplanetary magnetic field and the auroral zones, *Physical Review Letters* 6, 47-48, 1961.

14. Magnetohydrodynamic theory incorporates the effects of magnetic fields in the hydrodynamic description of ionized gases (plasmas). Hannes Alfvén, whose work profoundly influenced space physics, was awarded the 1970 Nobel Prize in physics "for fundamental work and discoveries in magnetohydrodynamics with fruitful applications in different parts of plasma physics."

15. National Security Space Architect, *Space Weather Architecture Study Transition Strategy*, March 1999. Available online at <http://schnarff.com/SpaceWeather/PDF/Reports/P-IIB/02.pdf>.

16. The organization of plasmas into large- and small-scale structures separated by thin boundaries has given rise to the picture, associated with Hannes Alfvén, of the "cellular structure" of solar system and astrophysical plasmas. Cf. H. Alfvén, *Cosmic Plasma*, D. Reidel, Dordrecht, The Netherlands, 1981.

2
Integrated Research Strategy for Solar and Space Physics

In developing a research strategy for solar and space physics for the coming decade, the committee was guided by several considerations. First, solar and space physics will achieve the greatest gains in understanding through coordinated investigations of its objects of study as interacting parts of complex systems. Second, to address the scientific challenges presented in Chapter 1, a combination of observational programs and complementary theory and modeling initiatives is needed, with the observational programs including both ground- and space-based elements. (The committee noted that many such efforts are already important components of planned or proposed agency programs.) Third, the vulnerability of society's technological infrastructure to space weather necessitates a mix of basic, targeted basic, and applied research initiatives that will lead both to advances in fundamental scientific knowledge and to progress in the application of that knowledge to the mitigation of space weather effects on technology and society. The research strategy that the committee has developed is thus an *integrated* one, a strategy that provides for the coordinated investigation of solar system plasmas as complex, coupled systems and that seeks to maximize the synergy between observational and theoretical initiatives and between basic research and targeted research programs. A final, critical consideration was cost realism. Accordingly, the committee exercised great care to ensure that its recommended research strategy is consistent with the anticipated budgets of the various federal agencies.

The programs and initiatives that constitute the committee's recommended strategy for solar and space physics research during the decade 2003-2013 are described in the sections that follow. The discussion is organized in terms of the scientific challenges set forth in the preceding chapter and emphasizes the complementarity of the various recommended initiatives. The section "Roadmap to Understanding" describes the criteria that the committee used and the decision-making process that it followed in establishing priorities among the programs and initiatives recommended by the disciplinary study panels. Cost estimates are presented in tabular form, and costs and phasing for all recommended programs are illustrated graphically in three waterfall charts.

THE SUN'S DYNAMIC INTERIOR AND CORONA

Helioseismological studies of the solar interior have attained a high degree of sophistication through the ground- and space-based measurements of Doppler shifts by GONG and SOHO, from which images of the magnetic fields and flow systems below the solar surface are deduced. Similarly, imaging and spectral data from the solar corona and transition region provided by the SOHO and TRACE satellites have demonstrated the central role of the magnetic field in controlling coronal dynamics. However, the mechanisms by which the solar magnetic field is generated, including its reversals and temporal cycles, are still not completely understood. Solar B, a mission now in development by Japan's Institute of Space and Astronautical Science (ISAS) and in which NASA is a participant, will measure how magnetic fields emerge onto the solar surface and reveal the details of the interaction between convective flows and magnetic fields. Also under development now is STEREO, which will measure the three-dimensional development and propagation of coronal mass ejections (CMEs) through the inner heliosphere.

The next important scientific steps in solar physics are the following: (1) to refine and sharpen the probing of the solar interior; (2) to treat the outer layers of the Sun and its atmosphere as a single system; (3) to develop the science of coronal mapping for measurement of the structure and strength of the coronal magnetic field; and (4) to make precise spectral measurements of the solar atmosphere over a broad range to map velocity distributions and to determine what spectral bands have the strongest effects at Earth and how they vary over the solar cycle. Three new programs, working in concert, will take these steps: the Solar Dynamics Observatory (SDO), the Advanced Technology Solar Telescope (ATST), and the Frequency-Agile Solar Radiotelescope (FASR).

SDO, which is part of the approved NASA Living With a Star program and is now in development, will explore the Sun from its center to the subsurface layers of the convection zone to the outer solar atmosphere, probing the subsurface origin of active regions with acoustic imaging of the convection zone and tracking their development in space and time. By virtue of the high data rate available from its geosynchronous orbit, SDO will make full-disk, high-resolution (in both space and time) maps of the Doppler velocities and magnetic fields of the Sun, making possible the exploration of the complete life cycle of active regions. SDO will also carry an ultraviolet (UV) spectrograph in order to understand the link between solar activity and solar spectral radiance.

The ground-based ATST, which is currently undergoing a design study funded by the NSF, will be a 4-meter facility that employs adaptive optics to study the solar magnetic field, from the photosphere up through the corona. In the lower atmosphere the telescope will achieve a flux density sensitivity of a few gauss or smaller. It will provide observations of the solar atmosphere at a high temporal cadence and with better than 0.1-arcsecond resolution, which is sufficient to resolve the pressure scale height and the photon mean free path in the solar atmosphere. ATST will thus enable critical tests of models of solar plasma processes.

An important ground-based element of the NSF-supported program recommended in this report is FASR (see Figure 2.1), which, like ATST, is now in a design study phase.[1] FASR represents a significant advance beyond existing solar radio instruments yet is well within the reach of emerging technologies. A wide range of solar features from within a few hundred kilometers of the visible surface of the Sun to high up in the solar corona can be studied in detail with the unique diagnostics available in the radio regime. FASR capabilities will include measuring the properties of both thermal and nonthermal electrons accelerated in solar flares, measuring coronal magnetic field strengths in active regions, and mapping kinetic electron temperatures throughout the chromosphere and corona.

Two of the major mysteries in solar physics are the fact that the Sun's corona is several hundred times hotter than the underlying photosphere and the fact that the coronal gases are accelerated to supersonic velocities within a few solar radii of the surface to form the solar wind. Resolving these mysteries—understanding how the corona is heated and how the solar wind originates and evolves in the inner heliosphere—has been identified by the Panel on the Sun and Heliospheric Physics as its top science priority for the coming decade. To answer these questions requires measurements from a spacecraft that passes as close to the solar surface as possible. A Solar Probe mission will make in situ measurements of the plasma, energetic particles, magnetic field, and waves inward of ~0.3 AU to an altitude of 3 solar radii above the Sun's surface. This region is one of the last unexplored frontiers in the solar system. Such measurements will locate the source and trace the flow of energy that heats the corona; determine the acceleration processes and find the source regions of the fast and

FIGURE 2.1 Artist's conception showing a portion of the Frequency-Agile Solar Radiotelescope's ~100-dish antenna array. FASR, which is the committee's highest-priority small initiative for solar physics, will produce high-resolution images of the Sun's atmosphere from the chromosphere up into the mid-corona. Courtesy of D.E. Gary (New Jersey Institute of Technology).

---

slow solar wind; identify the acceleration mechanisms and locate the source regions of solar energetic particles; and determine how the solar wind evolves with distance in the inner heliosphere. In addition, if suitable remote-sensing instruments are included, a Solar Probe mission can complement the in situ measurements with valuable close-up views of the Sun.[2] Because of the profound importance of the scientific questions that a Solar Probe will address, the committee recommends that this mission be implemented as soon as possible.

THE HELIOSPHERE AND ITS COMPONENTS

The heliosphere begins in the outer solar corona and ends at its interface with the interstellar medium (see Chapter 1). As such, it encompasses the entire solar system and is the domain of solar plasmas, magnetic fields, and energetic particles as well as interstellar dust, neutral atoms, and pickup ions. The plasma continuously flowing from the Sun in fast and slow solar winds is highly dynamic and turbulent, and the embedded CMEs and interplanetary shocks cause the impulsive transfer of solar energy to the magnetospheres (whether intrinsic or induced) of planets and small bodies. The heliosphere contains the connective tissue of the Sun-Earth connection; however, little is known about its source or its destiny. The new frontiers of heliospheric research lie at its inner boundary in the solar corona and its outer boundary with the interstellar medium. A Solar Probe mission will investigate the innermost boundary of the heliosphere. Moving outward from the Sun, ESA's Solar Orbiter mission will periodically corotate with the Sun in an elliptical orbit with perihelion of 45 solar radii. With its payload of imaging and in situ instruments, some of which will be contributed by the United States, Solar Orbiter will be poised to reveal the magnetic structure and evolution of the corona and the resulting effects on plasmas, fields, and energetic particles in the inner heliosphere. Participation in this mission will provide the United States with a highly leveraged means of investigating the structure and evolution of the inner heliosphere for the first time.

STEREO will perform stereoscopic imaging and two-point, in situ measurement of CMEs in the inner heliosphere. The next important step in understanding the heliospheric propagation of CMEs will be accomplished with a Multispacecraft Heliospheric Mission (MHM).[3] MHM will consist of four or more spacecraft separated in solar longitude and radius with at least one orbital perihelion at or within ~0.5 AU. Orbiting in and near the ecliptic plane, these spacecraft will make in situ measurements of plasmas, fields, waves, and energetic particles in the inner heliosphere, providing two-dimensional slices through propagating CMEs and the ambient solar wind.

The boundary between the solar wind and the local interstellar medium (LISM) is one of the last unexplored regions of the heliosphere. Very little is currently known about this boundary or the nature of the LISM that lies beyond it. The outer boundary of the heliosphere will eventually be sampled directly by an Interstellar Probe mission. Advances in propulsion technology are expected to make such a mission feasible during the decade 2010-2020. Although it cannot yet be included in the program recommended by the committee, an Interstellar Probe is a high-priority future mission for which the required technology investments should begin as soon as possible. In the meantime, certain aspects of the heliospheric boundary and the LISM can be studied by a combination of remote sensing and in situ sampling techniques. This investigation could be accomplished by an Interstellar Sampler mission traveling to distances of several AU to measure the neutral atoms of the LISM that penetrate well into the heliosphere and to obtain energetic neutral atom images and extreme ultraviolet images of the heliospheric boundary. Such a mission is gauged to be feasible within the resources of the Explorer program and so is not prioritized separately in this report.

SPACE ENVIRONMENTS OF EARTH AND OTHER SOLAR SYSTEM BODIES

Earth's magnetosphere and ionosphere formed the historical starting point for space physics research and remain an important focus for study because they constitute the human space environment and because they provide important prototypes for understanding the magnetospheres and ionospheres of other planets and small solar system bodies. In addition, the basic physical phenomena of space plasmas, which can be studied directly in Earth's magnetosphere, occur in remote and therefore inaccessible locations in the universe.

Having been the focus of numerous space- and ground-based investigations over the past four decades, the study of geospace has reached a level of maturity that brings the shortcomings of understanding into sharp focus. What specific physical processes transfer energy from the solar wind to geospace? What is the nature of the global response of Earth's magnetosphere and ionosphere to the variable solar-wind input? These are the most basic and important questions that can be asked about geospace, but they have not received satisfactory answers. However, the maturity of the field now allows the interrogation of large databases, the development of sophisticated models, and the construction of new, definitive experiments both for Earth's space environment and for that of other solar system bodies.

Magnetic fields are continually being created by the solar dynamo but are also continually being annihilated by both small- and large-scale magnetic reconnection in the corona. Reconnection converts magnetic energy to particle kinetic energy and heat, and the results are heating of the corona and explosive outbursts of solar flares. Similarly at Earth, magnetic fields are continually being created by the internal dynamo and being annihilated by reconnection with solar wind magnetic field lines at the dayside magnetopause and reconnection of open field lines within the geomagnetic tail. As at the Sun, these processes result in the energization of charged particles. NASA's Solar Terrestrial Probe (STP) mission Magnetospheric Multiscale (MMS) is designed to probe the reconnection process at the magnetopause and in the tail with a cluster of four spacecraft. MMS will benefit greatly from the groundbreaking research on magnetospheric and solar wind plasma dynamics that is being done with the European Cluster 2 mission. With its ability to adjust orbits and spacecraft separations, the MMS cluster will probe the boundary regions where reconnection is occurring and test directly theories of reconnection from the magnetohydrodynamic scale (thousands of kilometers) down to the ion and electron kinetic scales (kilometers).

Earth's ionosphere-thermosphere system is the site of complex electrodynamic processes that redistribute and dissipate energy delivered from the magnetosphere in the form of imposed electric fields and precipitating charged particles. Previous studies have revealed much about the composition and chemistry of this region and about its structure, energetics, and dynamics. However, a quantitative understanding has proved elusive because of the inability to distinguish between temporal and spatial variations, to resolve the variety of spatial and temporal scales on which key processes occur, and to establish the cross-scale relationships among small-, intermediate-, and large-scale phenomena. Geospace Electrodynamic Connections (GEC) (see Figure 2.2) is a multispacecraft STP mission that has been specifically designed to overcome these difficulties and to provide new physical insight into the coupling among the ionosphere, thermosphere, and magnetosphere.

Sounding rockets are an important part of NASA's Suborbital Program and have been a mainstay for the investigation of important small-scale physical processes in the ionosphere-thermosphere, for a wide range of magnetospheric studies, and for the development of new instruments for space physics. While the flight time of these rockets is small, their slow velocity through specific regions of space (nearly an order of magnitude less than that of orbiting vehicles) and their ability to sustain very high telemetry rates from multiple payloads launched from a single vehicle make them extremely useful for studying the fine structure of dynamic phenomena like the aurora. Moreover, some important regions of space are too low in altitude to be sampled by satellites (i.e., the mesosphere below 120 km), so sounding rockets are the only platforms from which direct in situ measurements can be carried out in these regions.

FIGURE 2.2 The Geospace Electrodynamic Connections (GEC) mission is a multispacecraft Solar Terrestrial Probe mission designed to study the multiple spatial and temporal scales on which the ionosphere-thermosphere system receives electromagnetic energy from the magnetosphere and redistributes and dissipates it through ion-neutral interactions. During the 2-year mission, the spacecraft will perform several "deep dipping" excursions that will allow them to make in situ measurements down to altitudes as low as 130 kilometers. Courtesy of NASA Goddard Space Flight Center.

The Advanced Modular Incoherent Scatter Radar (AMISR) is a planned NSF program that will bring the observing power of a modern multi-instrument, ground-based observatory to a variety of geophysical locations chosen to optimize the benefit of the observations to specific scientific inquiry (Figure 2.3). In addition to its incoherent scatter radar, AMISR will host a variety of ground-based diagnostics that together will address key science questions about atmosphere-ionosphere-magnetosphere (AIM) interactions

FIGURE 2.3 The Advanced Modular Incoherent Scatter Radar (AMISR) is the committee's top-ranked small initiative for ground-based geospace research. It combines a powerful state-of-the-art incoherent scatter radar with supporting optical and radio instrumentation in a transportable format. This flexibility enables the AMISR to study a wide range of ionospheric phenomena at polar, auroral, equatorial, and mid-latitudes and to act in close conjunction with other ground-based, suborbital, and satellite investigations of the geospace environment. This artist's conception depicts the fast-steering, multibeam RAO probing the dynamic auroral environment at high latitudes. Courtesy of J. Kelly and C.J. Heinselman (SRI International).

that can only be tackled with a detailed knowledge of evolving time and altitude variations in a specific region.

A Small Instrument Distributed Ground-Based Network[4] will combine state-of-the-art instrumentation with real-time communications technology to provide both broad coverage and fine-scale spatial and temporal resolution of upper atmospheric processes crucial to understanding the coupled AIM system. Placing a complement of instruments, including Global Positioning System receivers and magnetometers, at educational institutions will provide a rich, hands-on environment for students, while instrument clusters at remote locations will contribute important global coverage. This flexible NSF initiative will provide the simultaneous real-time measurements needed for assimilation into physics-based models and to address the space weather processes and effects in the upper atmosphere. These detailed, distributed measurements will complement the capabilities at the larger ground-based facilities that host the incoherent scatter radars.

Global magnetospheric imaging, currently available from the IMAGE mission, needs to be developed further, specifically with stereo imaging, which will be implemented for 1- to 30-keV neutral atoms with the Two Wide-Angle Imaging Neutral-Atom Spectrometers (TWINS) mission (launch dates in 2003 and 2004). Stereo imaging is essential because magnetospheric plasmas are optically thin. The Stereo Magnetospheric Imager (SMI), a proposed STP mission,[5] will obtain EUV images of the plasmasphere and neutral atom images of the plasma sheet and ring current from two widely spaced spacecraft, thereby greatly improving the accuracy with which global ion images can be produced. In addition, SMI will provide an auroral imaging capability, which has become a standard tool for assessing the global activity of the magnetosphere.

A large database of spacecraft magnetic field measurements has made it possible to construct good models of the average configuration of Earth's magnetosphere for different levels of magnetic activity. However, the actual instantaneous configuration of the magnetosphere is not expected ever to resemble any of its average states. What is needed is a dynamic image of the magnetic fields and plasmas within the magnetosphere. Measurements by a large constellation of spacecraft carrying magnetometers and simple plasma instruments can yield dynamic "images" of large portions of the magnetosphere. Magnetospheric Constellation (MagCon) is an STP mission that is designed to investigate the dynamics and structure of Earth's near magnetotail by means of a constellation of 50 to 100 spacecraft.

The Explorer program has long provided the opportunity for targeted investigations, which can complement the larger initiatives recommended by the committee. However, the committee is concerned that the overall rate at which solar and space physics missions are undertaken is still rather low. A revitalized University-Class Explorer (UNEX) program would address this problem while allowing innovative small investigations to be conducted. However, the very existence of a UNEX program depends critically on low-cost access to space, which is discussed in Chapter 7.

Jupiter has a giant magnetosphere that has gross similarities with Earth's

FIGURE 2.4 Electrical currents aligned with Jupiter's strong magnetic field couple the moons Io, Ganymede, and Europa with Jupiter's high-latitude ionosphere. Bright auroral emissions observed at the footpoints of the magnetic flux tubes linking the moons to the ionosphere (inset) are the signature of this coupling, the physics of which is only partially understood. (The letters superposed on the auroral image in the inset identify the emission features associated with the footpoints of Io, Ganymede, and Europa.) A probe to study the still unexplored polar regions of Jupiter's magnetosphere will answer basic questions about the nature of electrodynamic coupling between the Jovian atmosphere and magnetosphere and about auroral acceleration in a magnetospheric environment much different from Earth's. Hubble Space Telescope image courtesy of J.T. Clarke (Boston University) and NASA/Space Telescope Science Institute. Artist's rendering of the Jovian inner magnetosphere courtesy of J.R. Spencer (Lowell Observatory). Reprinted by permission from *Nature* 415:997-999 and cover, copyright 2002, Macmillan Publishers Ltd.

magnetosphere (bow shock, magnetopause, magnetotail), but Jupiter's strong magnetic field and Io's abundant plasma source produce dramatic differences. Jupiter's high rotation rate and the strong volcanic mass loading from Io create an intense outward centrifugal force that pulls plasma and magnetic flux outward from the inner magnetosphere and stretches the magnetic field into a disk-like configuration. Previous flyby missions and the Galileo orbiter have explored only the equatorial region of the Jovian magnetosphere. To understand how the magnetospheric plasma couples with the low-altitude auroral regions, measurements in a polar orbit are needed. From an elliptical polar orbit, a Jupiter Polar Mission (JPM) will determine the relative contributions of planetary rotation and the solar wind to the energy budget of the Jovian magnetosphere (see Figure 2.4). It will determine how the plasma circulates in the magnetosphere and assess the role of Io's volcanism in providing the mass that drives this circulation process. JPM will identify the charged particles responsible for the Jovian aurora and determine how these particles become energized, and it will identify the electrodynamic processes that couple the Jovian moons to the planet's high-latitude ionosphere.

THE ROLE OF THEORY AND MODELING IN MISSIONS AND FUNDAMENTAL SPACE PLASMA PHYSICS

Over the past decade and more, theory and modeling have played an increasingly important role both in defining satellite missions and other programs and in interpreting data through the development of new physical models. The enhanced role of theory and modeling is a consequence of the development of powerful computational tools that have facilitated the exploration of the dynamics of complex nonlinear plasma systems at both large magnetohydrodynamic spatial scales and kinetic microscales. Before the advent of these tools it was not possible to study these dynamical processes through analytic techniques alone.

In the coming decade, the deployment of clusters of satellites and large arrays of ground-based instruments will provide a wealth of data over a very broad range of spatial scales. Theory and computational models will play a central role, hand in hand with data analysis, in integrating these data into first-principles models of plasma behavior. Examples of the catalyzing influence of theory and computation on the interpretation of data from observational assets are many. A case in point is recent research in the area of magnetic reconnection, where new theoretical developments have spurred the successful search for signatures of kinetic reconnection in satellite data.

Theory and modeling activities have further importance in the application of the results from solar and space physics to allied fields such as astrophysics and fusion energy sciences. The heliosphere is the space physicist's laboratory wherein a wide variety of plasma processes, parameters, and boundary conditions are encountered (cf. Chapter 4). Many of these phenomena can be sampled directly and the results applied to systems where direct measurements are either very difficult or altogether infeasible. The identification of the critical dimensionless parameters controlling plasma dynamics through analysis combined with state-of-the-art computation is central to the successful extrapolation to differing environments, where absolute parameters may be very different from those that apply to solar-system plasmas.

NASA's Sun-Earth Connection Theory program has been very successful in focusing critical-mass theory and modeling efforts on specific topics in space physics. The NSF has long encouraged and supported theoretical and modeling investigators through its grants program. Theoretical work provides the community with state-of-the-art computational models that are developed and utilized with support from all the funding agencies. This theoretical understanding is used extensively for interpreting individual measurements as well as for developing physics-based data assimilation procedures for diverse but coupled parameters.

In view of the strongly coupled nature of the Sun-heliosphere system and the complementary objectives of the solar and space physics programs of the different federal agencies, two interagency initiatives are being proposed by the committee. One of these—the Virtual Sun—will incorporate a systems-oriented approach to theory, modeling, and simulation that will ultimately provide continuous models from the solar interior to the outer heliosphere.[6] The Virtual Sun will be developed in a modular fashion by focused attacks on various physical components of the heliosphere and on cross-cutting physical problems. The solar dynamo and three-dimensional reconnection are areas ripe for near-term concentration because they complement the planned ground- and space-based measurement programs.

The Coupling Complexity research initiative[7] will address multiprocess coupling, nonlinearity, and multiscale and multiregional feedback in space physics. The program advocates both the development of coupled global models and the synergistic investigation of well-chosen, distinct theoretical problems. For major advances to be made in understanding coupling complexity in space physics, sophisticated computational tools, fundamental theoretical analysis, and state-of-the-art data analysis must all be integrated under a single umbrella program. Again, this initiative is motivated by the anticipated ground- and space-based measurements that will provide spatially distributed data that must be incorporated into a single understanding of the physical processes at work in different volumes of geospace.

SPACE WEATHER

Earth's space environment is subject to severe episodic changes that are correlated with specific heliospheric disturbances. Like terrestrial weather, severe space weather can have disruptive and even destructive effects that must be mitigated. Effective mitigation requires characterization of the geospace environment in both its quiescent and disturbed states, an understanding of the physical processes that are involved in disturbed conditions (e.g., the acceleration of radiation belt electrons during magnetic storms), and, ultimately, the ability to forecast space weather events accurately. As in the case of terrestrial meteorology, global measurements and large-scale numerical modeling are required. The geospace environment poses a particularly challenging problem because the magnetosphere is a vast, three-dimensional structure whose distant components can be coupled quickly and directly by plasma phenomena such as field-aligned currrents.

While focusing specifically on those regions and phenomena that most directly affect the technological infrastructure of modern society, space weather research aims at a basic physical understanding of the geospace environment. Despite its practical orientation and benefits, space weather research is thus to be understood as targeted basic research rather than applied research in the traditional sense. The multiagency National Space Weather Program (NSWP) has developed a sound framework for rapid progress in this area. This program, with NSF as the lead agency, has provided better understanding of the role of magnetosphere-ionosphere coupling processes in producing radio-wave scintillations, large magnetic-field changes, and other space weather effects observable on the ground. The NSWP has been successful by providing support for focused activities that harness scientific understanding of the geospace environment to provide better specifications and predictions. The recent implementation of NASA's Living With a Star (LWS) program has added a spaceflight segment and additional modeling activity to the national space weather effort.

The Solar Dynamics Observatory, the first LWS mission, will provide nearly continuous observations of the Sun from geosynchronous orbit, gathering data on solar variability and monitoring the occurrence of geoeffective solar activity. In addition, SDO will monitor EUV irradiance, allowing the fundamental sources of ionization and heating in the ionosphere and thermosphere to be specified. The STEREO mission will provide a unique perspective from which to observe CMEs that are directed toward Earth, while a mission at L1 (which could be provided by a NOAA contribution, as recommended in Chapter 5, or possibly by Triana) will provide critical measurements of the solar wind plasma and the interplanetary magnetic field once the operational lifetime of the Advanced Composition Explorer is over. Global images of the magnetosphere from SMI will allow the response of the magnetosphere to solar inputs to be related to ionospheric and thermospheric responses described by data from ground-based facilities such as AMISR and from the GEC mission.

Despite the wide variety of data that can be used to specify and predict space weather, critical gaps in understanding have been identified. Some of the strongest effects of severe magnetospheric storms are produced by radiation belt particles, which often appear spontaneously and without precursors. The important energization and transport processes for these particles are not understood, primarily because with single satellites, changes in the particle distribution functions and electric and magnetic fields in the inner magnetosphere are measured at satellite orbital periods rather than at particle drift periods. Multiple spacecraft are needed to describe more fully the inner magnetospheric particle and field environment on appropriate time scales. Similarly, multipoint measurements are also needed in the ionosphere, where global changes occur on time scales that are short compared with the orbital periods and on spatial scales that are smaller than the longitudinal orbit spacing of even low-altitude satellites. To address these needs, the LWS Geospace Network will contain both a radiation-belt component and an ionosphere-thermosphere component, with each component consisting of two spacecraft.

Ionospheric effects at equatorial, auroral, and middle latitudes constitute a second major category of space weather effects that must be better characterized and understood. Electric fields and particles couple the lower atmosphere and ionosphere to the disturbed magnetosphere above, and large-scale changes in total electron content and ionospheric scintillations can adversely affect communications and navigation systems. Focused research on the processes involved will be provided by AMISR and other high-resolution ground- and space-based facilities, while the Small Instrument Distributed Ground-Based Network will provide the real-time data and the spatial/temporal resolution needed to address and model these effects in the large-scale upper atmosphere system.

To provide more advanced warning of solar wind disturbances as well as to warn of approaching corotating streams that could be missed by a single Sun-aligned probe, the committee recommends a Solar Wind Sentinels (SWS) mission[8] as part of the future LWS program. Three spacecraft equipped with 100-m solar sails will surround the Earth-Sun line at 0.98 AU with separations in the 0.1-AU range. In addition to providing advanced warning, these spacecraft will provide measurements of the spatial and temporal structure of heliospheric phenomena such as CMEs, interplanetary shocks, and solar-wind streams as they propagate toward Earth. The MHM will also provide information on spatial gradients that may affect the geo-effectiveness of solar eruptions.

The continuing growth in the number of solar and space physics data sets and the need to use multiple data sets in characterizing and predicting the geospace environment require that the accessibility of the data and modeling tools be assured. The Solar and Space Physics Information System proposed by the Panel on Theory, Modeling, and Data Exploration is designed to fill this need. This system will assign the tasks of data validation, access, and delivery to experienced scientists and provide access to the latest interpretive models for all interested scientists. By originating this effort inside the science community that is generating the information, attention will be given to timely delivery of data products together with uncertainty estimates.

ROADMAP TO UNDERSTANDING

The committee was charged with recommending "a systems approach to theoretical, ground-based, and space-based research that encompasses the flight programs and focused campaigns of NASA, the ground-based and base research programs of NSF, and the complementary operational programs of other agencies such as NOAA, DOD, and DOE." To accomplish this task, the approaches put forward by the four technical study panels were integrated, and those projects with the highest scientific impact and, in some cases, the greatest potential societal benefit were considered further (the programs with the greatest potential societal benefit are generally those related to the LWS program or the NSWP). This selected group of planned and proposed activities makes up the vigorous and essential research program that is briefly described in the earlier sections of this chapter. However, decisions had to be made about what the optimum affordable sequence of science activities (all of them highly meritorious) is, about which programs need to be operational simultaneously, and about which programs have the highest priority in case budgetary limitations or other unforeseen circumstances limit the overall effort. The committee embarked on this decision process by first looking at the total cost of each individual program (or, in the case of level-of-effort activities such as the NASA Suborbital Program, the accumulated cost over the next decade) to categorize the activity as large (>$400 million), moderate ($250 million to $400 million), or small (<$250 million). By this measure, some relatively large NSF investments, such as AMISR and FASR, were grouped together with the small NASA programs.

The first consideration in developing a recommended program was to map the scientific challenges (Chapter 1) to the objectives and capabilities of the planned and proposed program elements. Table 2.1 shows the mapping between the science challenges and the various missions and facilities selected by the committee.

The underlying vitality of the solar and space physics discipline depends heavily on the robustness of NASA's Supporting Research and Technology (SR&T) program and of NSF's base science program and the CEDAR, GEM, and SHINE initiatives, as well as of the NSF-led NSWP. In addition, the development of a systems-level understanding of the geospace environment requires that a high priority be assigned to establishing an initiative to address the complex coupling processes between the different regions. Thus a separate "Vitality" category was identified, and its programs were prioritized in a manner similar to that for the programs in the large, moderate, and small categories. Under the rubric Vitality, the committee considered a number of existing and new activities that stabilize and enhance the connective fabric of the solar and space physics program. In addition to the aforementioned NSF base program, the NSWP, and NASA's SR&T program, NASA's Sun-Earth Connection Theory program and Guest Investigator program have been particularly important. Further support for these programs as well as support for new programs—the Coupling Complexity Research initiative, the Virtual Sun, the Solar and Space Physics Information System, and the LWS Data Analysis, Theory, and Modeling program—was also carefully considered.

Within the four main categories, the science impact of each program on the overall solar and space physics discipline was considered along with the program's potential societal benefits. The programs were ranked by a consensus of the members of the committee following extensive discussions over the course of two prioritization meetings attended by all committee members. The overall rankings are shown in Table 2.2, which also gives brief descriptions of the various programs and missions.

TABLE 2.1 Mapping of Missions and Facilities to Scientific Challenges

	Scientific Challenges				
Missions and Facilities	The Dynamic Solar Interior and Corona	The Heliosphere and Its Components	Earth and Planetary Space Environments	Fundamental Space Plasma Physics	Space Weather
Large					
Solar Probe	X	X		X	
Moderate					
Magnetospheric Multiscale			X	X	
Geospace Network			X		X
Jupiter Polar Mission			X	X	
Multispacecraft Heliospheric Mission		X		X	
Geospace Electrodynamic Connections			X	X	X
Suborbital Program	X		X	X	
Magnetospheric Constellation			X	X	X
Solar Wind Sentinels		X			X
Stereo Magnetospheric Imager			X		X
Small					
Frequency-Agile Solar Radiotelescope	X			X	X
Advanced Modular Incoherent Scatter Radar			X		X
L1 Monitor					X
Solar Orbiter	X	X			
Small Instrument Distributed Ground-Based Network			X		X
University-Class Explorers	X		X	X	X
Planned/approved initiatives					
Solar Dynamics Observatory	X				X
Advanced Technology Solar Telescope	X			X	X

Recommendation: The committee recommends the approval and funding of the prioritized programs listed in Table 2.2.

While all the programs in Table 2.2 are exceptionally meritorious, small discriminators emerged in the ranking process. The maturity of the study phase of the Magnetospheric Multiscale mission and its attack on the fundamental problems of magnetic reconnection are highly valued. So, too, are the fundamental science questions related to ionospheric variability and particle acceleration in the inner magnetosphere that will be addressed by the LWS Geospace Network missions. Together, these programs will significantly improve the nation's ability to specify and predict the response of the geospace environment to solar activity. Similarly, the ability to image the Sun with rapid temporal and spatial resolution using the Frequency-Agile Solar Radiotelescope, an important science opportunity in its own right, greatly complements the science program associated with the Solar Dynamics Observatory. Finally, the AMISR will allow many fundamental parameters to be measured in regions important to understanding the response of the ionosphere and atmosphere to external energy inputs driven by the Sun. In addition to their importance for ground-based basic research in solar and space physics, the committee also notes that FASR and AMISR will contribute importantly to the national space weather effort—FASR by providing real-time and near-real-time solar data and AMISR through improved understanding of disturbances of the upper atmosphere and ionosphere associated with space weather events.

Beyond these programs are others with great scientific merit and societal benefit that are directly related to an improved specification of how the geospace system responds to changes in inputs from the Sun and solar wind. An important element of this endeavor is the maintenance of an L1 solar-wind monitor, which the committee recommends be implemented by NOAA (cf. Chapter 5).

The rankings set forth in Table 2.2 were combined with available cost estimates and considerations of technical readiness to arrive at a phasing of programs that could be conducted in the next decade and remain within a reasonable budget. Table 2.3 shows the costing and readiness factors that the committee used to construct the implementation schedule for the NASA initiatives shown in Figure 2.5. In each case the costs include Phases B through D only. MMS and Geospace Network take their place as the important first steps in the committee's recommended program. These and the other flight programs are superposed on enhancements to the SR&T program and the Suborbital Program that reflect the high scientific produc-

TABLE 2.2 Priority Order of the Recommended Programs in Solar and Space Physics

Type of Program	Rank	Program	Description
Large	1	Solar Probe	Spacecraft to study the heating and acceleration of the solar wind through in situ measurements and some remote-sensing observations during one or more passes through the innermost region of the heliosphere (from ~0.3 AU to as close as 3 solar radii above the Sun's surface).
Moderate	1	Magnetospheric Multiscale	Four-spacecraft cluster to investigate magnetic reconnection, particle acceleration, and turbulence in magnetospheric boundary regions.
	2	Geospace Network	Two radiation-belt-mapping spacecraft and two ionospheric mapping spacecraft to determine the global response of geospace to solar storms.
	3	Jupiter Polar Mission	Polar-orbiting spacecraft to image the aurora, determine the electrodynamic properties of the Io flux tube, and identify magnetosphere-ionosphere coupling processes.
	4	Multispacecraft Heliospheric Mission	Four or more spacecraft with large separations in the ecliptic plane to determine the spatial structure and temporal evolution of coronal mass ejections (CMEs) and other solar-wind disturbances in the inner heliosphere.
	5	Geospace Electrodynamic Connections	Three to four spacecraft with propulsion for low-altitude excursions to investigate the coupling among the magnetosphere, the ionosphere, and the upper atmosphere.
	6	Suborbital Program	Sounding rockets, balloons, and aircraft to perform targeted studies of solar and space physics phenomena with advanced instrumentation.
	7	Magnetospheric Constellation	Fifty to a hundred nanosatellites to create dynamic images of magnetic fields and charged particles in the near magnetic tail of Earth.
	8	Solar Wind Sentinels	Three spacecraft with solar sails positioned at 0.98 AU to provide earlier warning than L1 monitors and to measure the spatial and temporal structure of CMEs, shocks, and solar-wind streams.
	9	Stereo Magnetospheric Imager	Two spacecraft providing stereo imaging of the plasmasphere, ring current, and radiation belts, along with multispectral imaging of the aurora.
Small	1	Frequency-Agile Solar Radiotelescope	Wide-frequency-range (0.3-30 GHz) radiotelescope for imaging of solar features from a few hundred kilometers above the visible surface to high in the corona.

TABLE 2.2 Continued

Type of Program	Rank	Program	Description
	2	Advanced Modular Incoherent Scatter Radar	Movable incoherent scatter radar with supporting optical and other ground-based instruments for continuous measurements of magnetosphere-ionosphere interactions.
	3	L1 Monitor	Continuation of solar-wind and interplanetary magnetic field monitoring for support of Earth-orbiting space physics missions. Recommended for implementation by NOAA.
	4	Solar Orbiter	U.S. instrument contributions to European Space Agency spacecraft that periodically corotates with the Sun at 45 solar radii to investigate the magnetic structure and evolution of the solar corona.
	5	Small Instrument Distributed Ground-Based Network	NSF program to provide global-scale ionospheric and upper atmospheric measurements for input to global physics-based models.
	6	University-Class Explorer	Revitalization of University-Class Explorer program for more frequent access to space for focused research projects.
Vitality	1	NASA Supporting Research and Technology	NASA research and analysis program.
	2	National Space Weather Program	Multiagency program led by the NSF to support focused activities that will improve scientific understanding of geospace in order to provide better specifications and predictions.
	3	Coupling Complexity	NASA/NSF theory and modeling program to address multiprocess coupling, nonlinearity, and multiscale and multiregional feedback.
	4	Solar and Space Physics Information System	Multiagency program for integration of multiple data sets and models in a system accessible by the entire solar and space physics community.
	5	Guest Investigator Program	NASA program for broadening the participation of solar and space physicists in space missions.
	6	Sun-Earth Connection Theory and LWS Data Analysis, Theory, and Modeling Programs	NASA programs to provide long-term support to critical-mass groups involved in specific areas of basic and targeted basic research.
	7	Virtual Sun	Multiagency program to provide a systems-oriented approach to theory, modeling, and simulation that will ultimately provide continuous models from the solar interior to the outer heliosphere.

TABLE 2.3 Cost Estimates for Phases B Through D and Technical Concern Levels for the Recommended Flight Missions and Ground-Based Facilities (million dollars)

Program	Cost (FY 2002)	Technical Concern
Solar Probe	650	Moderate-high
Geospace Electrodynamic Connections	300	Low
Geospace Network	400	Low
Jupiter Polar Mission	350	Moderate
Magnetospheric Constellation	325	High
Magnetospheric Multiscale	350	Low
Multi-Heliospheric Probes	300	Moderate
Solar Wind Sentinels	300	Moderate
Stereo Magnetospheric Imager	300	Low
Suborbital Program	30/yr (2002)–60/yr (2012)	Low
Frequency-Agile Solar Radiotelescope	60	Low
L1 Monitor	100	Low
Relocatable Atmospheric Observatory	65	Low
Small Instrument Distributed Ground-Based Network	5/yr	Low
Solar Orbiter	100	Moderate
University-Class Explorer	35/yr	Moderate

NOTE: Cost estimates were obtained directly from agency estimates whenever possible. Large, medium, and small programs are grouped separately in alphabetical order.

tivity of these efforts and the need to ensure that the overall science underpinning the major missions does not fall short of its full potential. Because the committee believes that it is imperative to understand the three-dimensional development of energetic solar events such as CMEs as they propagate from the inner heliosphere to 1 AU and beyond, it has included a Multispacecraft Heliospheric Mission in its recommended program. The committee is aware that the different ways in which a mission of this type might be implemented present different technology challenges. For example, an inner heliospheric mission might best be implemented using solar sails. However, in view of the importance of the understanding to be gained through multipoint measurements in the inner heliosphere, the committee recommends the early implementation of an MHM that is consistent with existing technology.

The committee emphasizes the scientific importance of investigating the complex space environments of other planets. Such investigations serve

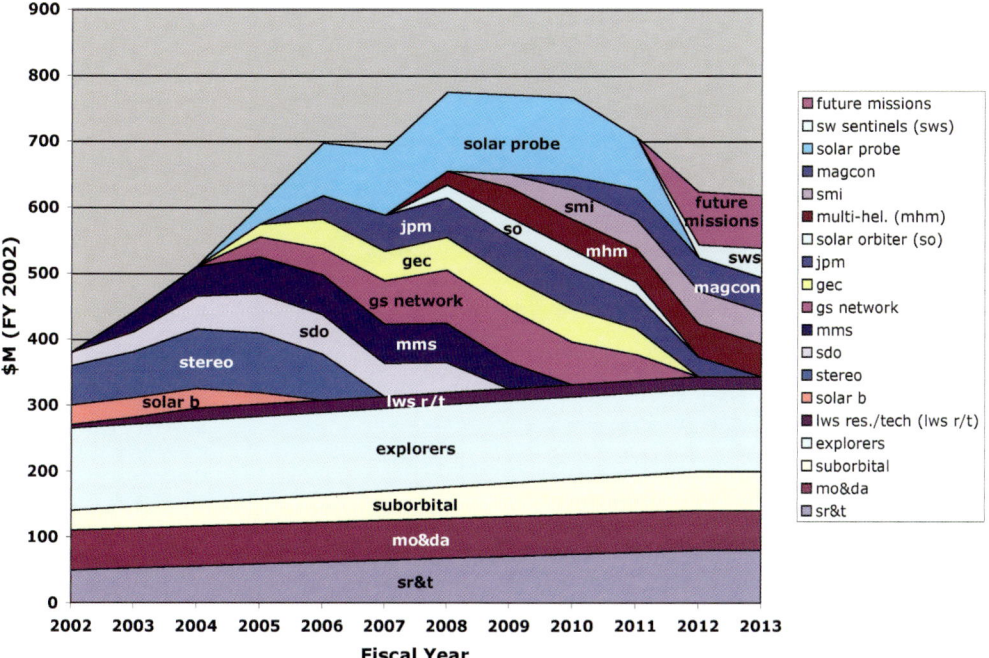

FIGURE 2.5 Recommended phasing of the highest-priority NASA missions, assuming an early implementation of a Solar Probe mission. Solar Probe was the Survey Committee's highest priority in the large mission category, and the committee recommends its implementation as soon as possible. However, the projected cost of Solar Probe is too high to fit within plausible budget and mission profiles for NASA's Sun-Earth Connection (SEC) Division. Thus, as shown in this figure, an early start for Solar Probe would require funding above the currently estimated SEC budget of $650 million per year for fiscal years 2006 and beyond. Note that mission operations and data analysis (MO&DA) costs for all missions are included in the MO&DA budget wedge.

as rigorous tests of the ideas developed from the study of Earth's own environment while extending the knowledge base to other solar-system bodies. The committee therefore strongly recommends a Jupiter Polar Mission, which will study energy transfer in a magnetosphere that is the largest object in the solar system and that, unlike Earth's, is powered principally by planetary rotation.

A Solar Probe is the only large mission considered by the committee for which the technical readiness is appropriate for implementation in the decade 2003-2013. The scientific merit of a Solar Probe mission (which was the top priority of the Panel on the Sun and Heliospheric Physics) is out-

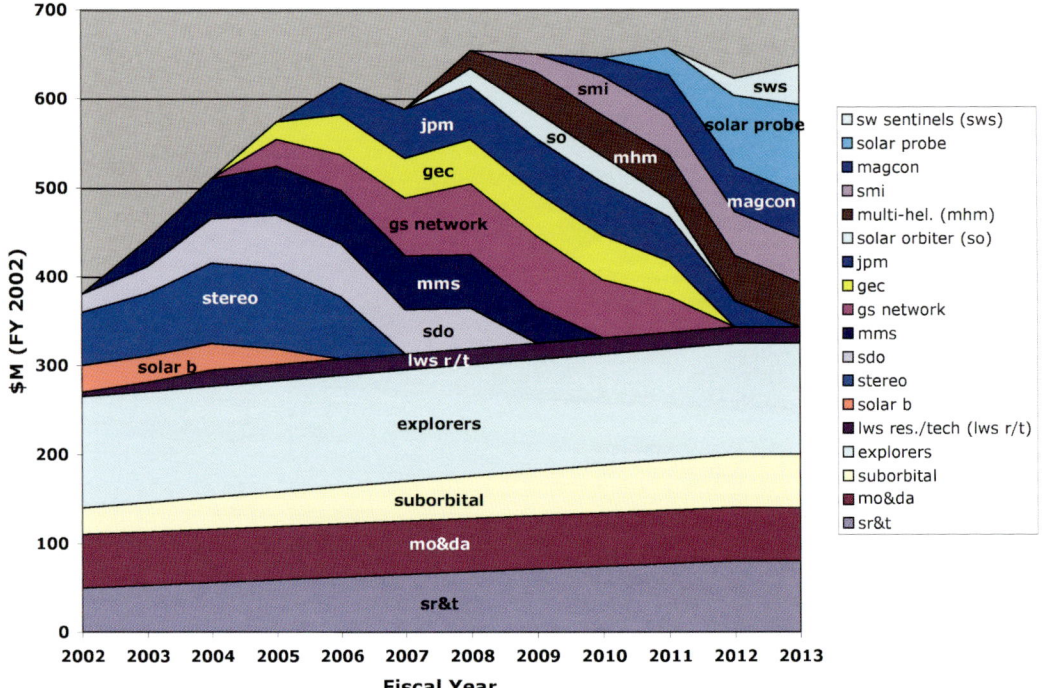

FIGURE 2.6 Recommended phasing of the highest-priority NASA missions if budget augmentation for Solar Probe is not obtained. MO&DA costs for all missions are included in the MO&DA budget wedge.

standing, and the committee recommends its implementation as soon as possible. However, the projected cost of such a mission is too high to fit within either the LWS program or the STP program. Thus, for this important program, separate funding would be required, as is illustrated in Figure 2.5, in which the Solar Probe budget profile extends above the projected Sun-Earth Connection (SEC) baseline budget projection.

In the event that funding augmentation for a Solar Probe mission cannot be secured, the recommended program can still be implemented but with Solar Probe having to start later, which would not be desirable or in keeping with its high scientific priority. Such an alternative sequencing is illustrated in Figure 2.6, which is based on a conservative estimate of the SEC budget.

A summary of the expected NSF funding profile and the recommended phasing of major NSF initiatives is shown in Figure 2.7. An important aspect of the recommended program is the additional funding for facilities

INTEGRATED RESEARCH STRATEGY

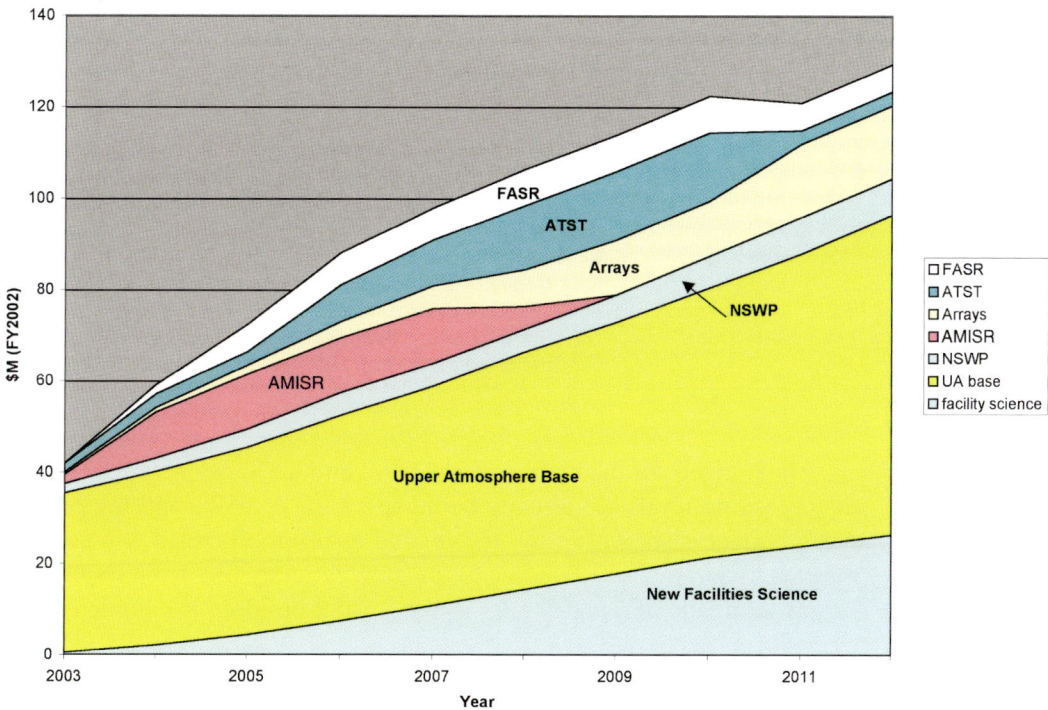

FIGURE 2.7 Recommended phasing of major new and enhanced NSF initiatives. The budget wedge for new facilities science refers to support for guest investigator and related programs that will maximize the science return of new ground facilities to the scientific community. Funding for new facilities science is budgeted at approximately 10 percent of the aggregate cost for new NSF facilities.

---

science ("guest investigator"), which will allow the scientific community to reap the scientific results that the investment in new ground-based observatories will make possible. The new facilities science is budgeted at ~10 percent of the aggregate cost of the new NSF facilities. The cost estimate for the NSF Upper Atmosphere base program includes costs for the CEDAR, GEM, and SHINE research initiatives, which coordinate community research activity and encourage strong student participation, as well as for individual research support in the areas of aeronomy, magnetospheric physics, and solar-terrestrial relations. NSF projections are that this baseline will double within 5 years; the committee has included this projection, doubling funding for the Upper Atmosphere base, the NSWP, and new facilities every 5 years.

TABLE 2.4 Deferred High-Priority Flight Missions (Listed Alphabetically)

Mission	Reason for Deferral
Large	
Interstellar Probe	Advanced propulsion technology needed
Moderate	
Auroral Cluster	Lower priority than moderate missions shown in Table 2.2
Dayside Boundary Constellation	Next step after Magnetospheric Constellation
Geospace System Response Imagers	Advanced solar sail technology needed
Io Electrodynamics	Next step after Jupiter Polar Mission
Mars Aeronomy Probe	Not supported by existing SEC mission lines
Reconnection and Microscale Probe	Lower priority than moderate missions shown in Table 2.2
Venus Aeronomy Probe	Not supported by existing SEC mission lines

The budgets for the highly rated AMISR and FASR projects begin immediately (some funds are already being spent for AMISR), and the ongoing ATST budget is shown on the reasonable assumption that it will continue through to completion. Funding for the new program involving small arrays begins gradually and accelerates as the AMISR buildout is completed.

DEFERRED HIGH-PRIORITY FLIGHT MISSIONS

Several large and moderate missions suggested by the panels were given high priority by the committee but were not included in the recommended program because of the overall budget constraint, mission sequencing requirements, or technical readiness issues. These missions are listed alphabetically in Table 2.4.

SUMMARY

The committee based the foregoing national strategy for the next decade of solar and space physics research on a systems approach to understanding the physics of the coupled solar-heliospheric environment. The work of the study panels was essential to the committee's deliberations and conclusions, as were all of the public outreach activities that were undertaken. The existence of ongoing NSF programs and facilities in solar and space physics, of two complementary mission lines in the NASA Sun-Earth Connection Division—Solar Terrestrial Probes (STP, basic research) and Living With a Star (LWS, targeted basic research)—and of applications and operations activities in NOAA and the DOD facilitated the development of an integrated research strategy.

As a key first element of its systems-oriented strategy, the committee endorsed three approved NASA missions, Solar-B (STP), STEREO (STP), and the Solar Dynamics Observatory (LWS). Together with ongoing NSF-supported solar physics programs and facilities as well as the start of the Advanced Technology Solar Telescope (ATST), these missions constitute a synergistic approach to the study of the inner heliosphere. This approach will involve coordinated observations of the solar interior and atmosphere and of the formation, release, evolution, and propagation of CMEs toward Earth. Later in the decade covered by the study, an overlapping of the Solar Dynamics Observatory (LWS), the ATST, and the Magnetospheric Multiscale mission (STP), together with the start of the Frequency-Agile Solar Radiotelescope, will form the intellectual basis for a comprehensive investigation of magnetic reconnection in the dense plasma of the solar atmosphere and the tenuous plasmas of geospace.

The committee's ranking of the Geospace Electrodynamic Connections (STP) and Geospace Network (LWS) missions acknowledges the importance of studying Earth's ionosphere and inner magnetosphere as a coupled system. Together with a ramping up of the launch opportunities in the Suborbital Program and the implementation of both the AMISR and the Small Instrument Distributed Ground-Based Network, these missions will provide a unique opportunity to study the local electrodynamics of the ionosphere down to altitudes where energy is transferred between the magnetosphere and the atmosphere, while simultaneously investigating the global dynamics of the ionosphere and radiation belts. The implementation of the L1 Monitor (NOAA) and of the Vitality programs will be essential to the success of this systems approach to basic research and targeted basic research. Later on in the committee's recommended program, concurrent operations of a Multispacecraft Heliospheric Mission (LWS), Stereo Magnetospheric Imager (STP), and MagCon (STP) will provide opportunities for a coordinated approach to understanding the large-scale dynamics of the inner heliosphere and Earth's magnetosphere (again with strong contributions from the ongoing and new NSF initiatives).

To understand the genesis of the heliospheric system it is necessary to determine the mechanisms by which the solar corona is heated and the solar wind is accelerated and to understand how the solar wind evolves in the innermost heliosphere. These objectives will be addressed by a Solar Probe mission. Because of the importance of these objectives for the overall understanding of the solar-heliosphere system, as well as of other stellar systems, a Solar Probe mission should be implemented as soon as possible within the coming decade. Solar Probe measurements will be complemented by correlative observations from such initiatives as Solar Orbiter, the Solar Dynamics Observatory, the Advanced Technology Solar Telescope, and the Frequency-Agile Solar Radiotelescope.

Similarly, because comparative magnetospheric studies are so important for advancing understanding of basic magnetospheric processes, the committee has assigned high priority to a space physics mission to study high-latitude electrodynamic coupling at Jupiter. Such a mission will provide both a means of testing and refining theoretical concepts developed largely in studies of the terrestrial magnetosphere and a means of studying in situ the electromagnetic redistribution of angular momentum in a rapidly rotating system, with results relevant to such astrophysical questions as the formation of protostars.

NOTES

1. While both FASR and ATST are now in the design phase, ATST is the more mature of the two programs. (The FASR design study began only recently, after the Panel on the Sun and Heliospheric Physics had finalized its recommendations.) The committee and the panel therefore regarded ATST as an "approved initiative" and considered FASR to be a "new initiative" that was to be evaluated and ranked with other new initiatives. The committee considers both ATST and SDO to be essential elements of a baseline program on which the recommended new initiatives will build. It recommends continued funding for technology development in support of ATST (cf. Chapter 3 of this report).

2. As explained in note 1 in the Executive Summary, the Solar Probe mission recommended by the Panel on the Sun and Heliospheric Physics emphasizes in situ measurement of the solar wind and corona near the Sun. The panel does not consider remote sensing a top priority on a first mission to the near-Sun region, although it does allow as a possible secondary objective remote sensing of the photospheric magnetic field in the polar regions. While accepting the panel's assessment of the critical importance of the in situ measurements for understanding coronal heating and solar wind acceleration, the committee does not wish to rule out the possibility that some additional remote-sensing capabilities, beyond the remote-sensing experiment to measure the polar photospheric magnetic field envisioned by the panel, can be accommodated on a Solar Probe within the cost cap set by the committee.

3. For more details on MHM, see the report of the Panel on the Sun and Heliospheric Physics in the companion volume to this report (2003, in press).

4. Cf. the report of the Panel on Atmosphere-Ionosphere-Magnetosphere Interactions in the companion volume to this report (2003, in press).

5. Described in the *Sun-Earth Connection Roadmap: Strategic Planning for the Years 2000-2020*, NASA Office of Space Science, 1997.

6. Cf. the report of the Panel on the Sun and Heliospheric Physics in the companion volume to this report ((2003, in press).

7. Cf. the report by the Panel on Theory, Modeling, and Data Exploration in the companion volume to this report ((2003, in press).

8. Described in the *Sun-Earth Connection Roadmap: Strategic Planning for the Years 2000-2020*, NASA Office of Space Science, 1997.

3

Technology Development

Historically, the fields of solar and space physics were founded on ground-based observation of the Sun and the near-Earth space environment. Great advances occurred during and after the International Geophysical Year 1957-1958, when globally coordinated, ground-based observations and measurements made by balloons, sounding rockets, and orbiting spacecraft became available. Many of the most fundamental discoveries in solar and space physics were made with these early systems, which employed "borrowed" technology such as captured V2 rockets and intermediate-range ballistic missiles. As time went by, dramatic advances in spacecraft technology, and particularly in scientific instrumentation, led to the rapid and complete reconnaissance of geospace and the identification of dynamic features of the solar corona that are closely connected with geomagnetic activity. Studies of coronal activity benefited greatly from the deployment of solar imagers on the Skylab space station. In 1973-1974, for example, these imagers verified and extended earlier rocket measurements identifying coronal holes as the source of high-speed solar wind streams. Rapid development of both ground- and space-based solar imagers has enabled multispectral imaging of the Sun with ever higher resolution, now in the sub-arcsecond range. These observations, together with theory and numerical simulations, have come closer than ever to revealing how electromagnetic energy is explosively converted to plasma kinetic and heat energy in the solar corona.

During the more than four decades of the space age, the magnetic and gaseous environments of eight of the planets and of several planetary satellites and comets, along with different regions of the heliosphere, have been explored to varying degrees. Prior to the first dedicated dual-spacecraft mission in 1977, simultaneous measurements at several locations in geospace had been achieved on a hit-or-miss basis since the mid-1960s. The ability to make more than single-point measurements enabled researchers to establish the characteristic structure and motions of various plasma boundaries and shocks in geospace. Understanding of the global dynamics of Earth's magnetosphere was spurred by the advent of global auroral imaging in 1981 and more recently (2000) by the comprehensive imaging of inner magnetospheric plasmas. Targeted in the future is the three-dimensional dynamic "imaging" of magnetic fields and charged particle populations in large parts of the magnetosphere using constellations of many tiny spacecraft.

Theoretical understanding of solar and space physics has been aided greatly by the rapid development of computing technology, which has allowed investigations of the global magnetohydrodynamic behavior of space plasmas as well as of their kinetic behavior on a local scale. Large data sets are being collected, and the assimilation of these data into physics-based models is an important intermediate goal in the search for an ultimate predictive understanding of planetary space environments.

While the above-mentioned "borrowed" technologies have made possible many of the advances in solar and space physics, the development of new technologies is needed to enable the solar and space physics research communities to address some of the key scientific questions set forth in Chapter 1. The committee has identified seven main areas in which focused technology development, based on the immediate and projected needs of solar and space physics research, is required:

- Sending spacecraft to the planets and beyond as efficiently as possible;
- Developing highly miniaturized spacecraft and advanced spacecraft subsystems for missions involving constellations of multiple spacecraft;
- Developing highly miniaturized sensors of charged and neutral particles and photons;
- Gathering and assimilating the data from multiple platforms;
- Integrating large space-physics databases into physics-based numerical models;
- Deploying reliable, unmanned, ground-based ionospheric and geomagnetic measurement stations; and
- Developing a high-resolution, ground-based solar imager.

Progress in each of these areas will require (1) recognition of the limitations of present technologies; (2) development of new technologies such as in-space propulsion, miniaturization, and low-power techniques for spacecraft systems and more capable ground-based instrument arrays; and (3) investment in existing technological capabilities to maximize productivity and efficiency. The committee expands on each of these issues below.

TRAVELING TO THE PLANETS AND BEYOND

New propulsion and power technologies are needed to enable further planetary and heliospheric missions. An important part of any space mission is the ability to loft a spacecraft into space and propel it to its intended orbit or destination. At present this task is achieved by low- or moderate-technology propulsion systems operated by highly sophisticated avionics. Limited options are provided by the fleet of launch vehicles in the United States, and the frequency of their use for solar and space physics is insufficient to reduce costs below the levels that the solar and space physics research community has come to accept. There is also no new, groundbreaking technology that promises to change this constraint. It should be recognized, therefore, that for small- and moderate-class missions, launch costs will remain a significant fraction of the overall mission costs. However, as discussed in Chapter 7, there is a new DOD program to develop launch systems that could potentially provide very-low-cost access to low Earth orbit for small (75-100 kg) spacecraft.

Near-Earth missions slated for launch in the relatively near term will use hydrazine thrusters to provide impulsive changes in velocity for station keeping—required, for example, for magnetospheric multiprobe missions—and for excursions to low altitudes by the Geospace Electrodynamic Connections mission. However, many future missions require satellite trajectories that are not possible without continuous thrust capability. NASA has therefore created the In-Space Propulsion (ISP) program within the Office of Space Science to research and develop advanced propulsion and associated power systems that will supply the needed continuous thrust capability. Two propulsion technologies that are receiving particular attention within the ISP program are solar sails and nuclear electric propulsion.

Solar sails (see Figure 3.1) have long been envisioned as a simple, inexpensive means of propulsion that could provide access to and maintenance of unstable orbits that would otherwise require large, expensive propulsion systems. The potential of solar sails is being explored for a number of space physics missions proposed or now under study, including Solar Wind Sentinels, ESA's Solar Orbiter, the Solar Polar Imager, and the Interstellar Probe. The Planetary Society is sponsoring a solar-sail demonstration, with a target launch date of 2003. NASA's Office of Space Science has issued a solicitation for in-space propulsion technology (ROSS-2002)

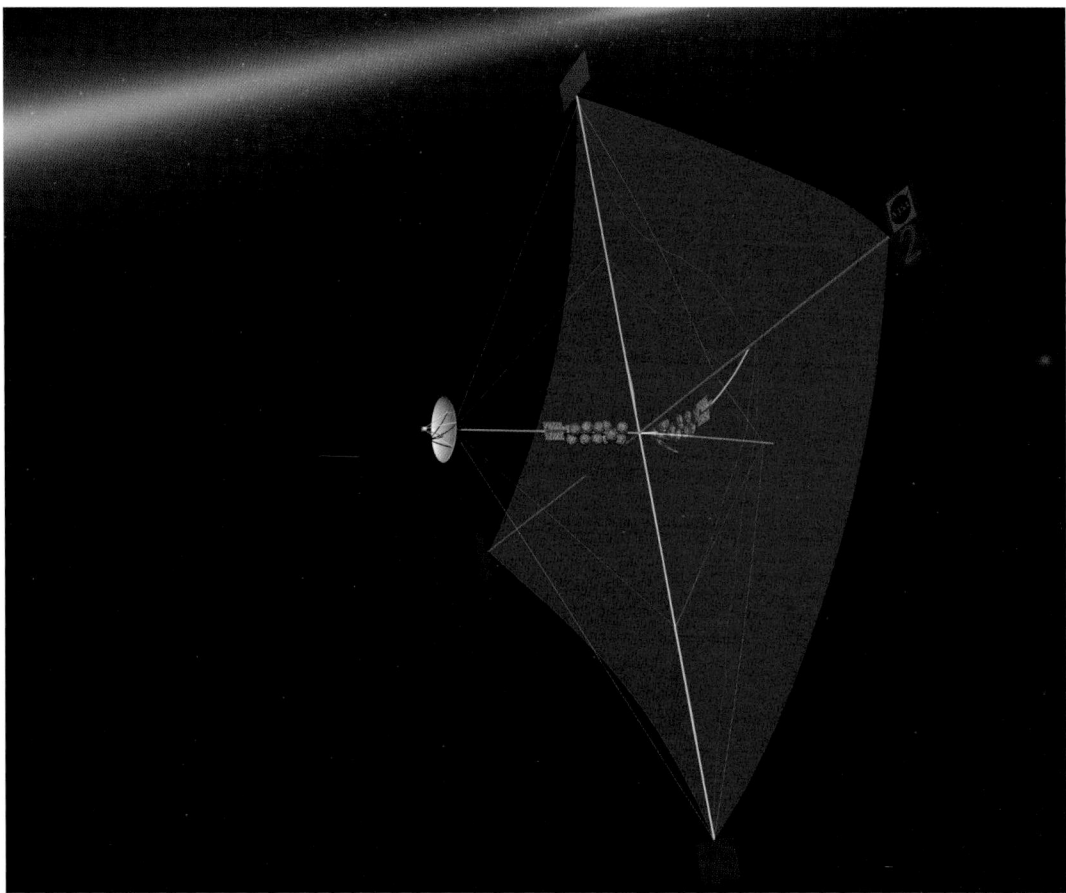

FIGURE 3.1 Artist's concept of a solar-sail-powered spacecraft. Solar sails, which use the radiation pressure of sunlight for propulsion or station keeping, are being considered for an Interstellar Probe and, within the Living With a Star program, for a multispacecraft "sentinel" mission to study the evolution of coronal mass ejection-driven disturbances and other solar wind structures as they propagate through the inner heliosphere. Courtesy of NASA Marshall Space Flight Center.

that includes a request for the development of an engineering model of a prototype solar sail system and for the development of a simulation package to validate sail performance. Because of the time required for development, testing, and validation, use of solar sails for actual solar and space physics missions is likely to be some 7 to 10 years in the future.

Missions to the outer planets and beyond, including an Interstellar Probe, require propulsion capabilities that significantly exceed those of the present fleet of launch vehicles. Nuclear electric propulsion (NEP) is being considered as a promising solution to the problem of providing the high-performance propulsion capabilities that such deep-space missions will need. The NASA 2003 budget submitted to Congress contains funding to support the study of NEP technology, and the ROSS-2002 NASA research announcement solicits proposals for propulsion and power studies that will lead eventually to the development of an NEP capability. When such a propulsion technology becomes available, there is little doubt that an Interstellar Probe will become one of the most exciting of the future missions that could reach beyond the solar system.

Although recent advances in high-efficiency solar arrays can support small- and moderate-scale missions to Jupiter, radioisotope thermoelectric generators (RTGs) are needed for most missions to the outer solar system. They might also simplify the design of a Solar Probe mission and make multiple passes into ~4 solar radii a possibility. Pioneers 10 and 11, Voyagers 1 and 2, and the Ulysses, Galileo, and Cassini missions have all used RTGs as the source of their electrical power, with the Voyager and Pioneer missions having demonstrated long-term (>25-year) reliability. The 2003 NASA budget provides funding for the renewed production of RTGs.

Recommendation: NASA should assign high priority to the development of advanced propulsion and power technologies required for the exploration of the outer planets, the inner and outer heliosphere, and the local interstellar medium. Such technologies include solar sails, space nuclear power systems, and high-efficiency solar arrays. Equally high priority should be given to the development of lower-cost launch vehicles for Explorer-class missions and to the reopening of the radioisotope thermoelectric generator (RTG) production line.

These advanced technologies include solar sails, space nuclear power systems, and high-efficiency solar arrays. Equally high priority should be given to the development of lower-cost launch vehicles for Explorer-class missions and to reopening the RTG production line.

ADVANCED SPACECRAFT SYSTEMS

The development of highly miniaturized spacecraft and advanced spacecraft subsystems is needed to enable planned and proposed missions that involve constellations of multiple spacecraft. NASA has studied several science missions that require multiple distributed satellites, including the Magnetospheric Constellation (MagCon) mission. MagCon is an example of a science mission that is driving "nanosatellite" technology development. Meeting the objectives of this mission will require multiple simultaneous in situ observations from 50 to 100 spacecraft, which in turn is driving technology on many fronts. Especially important is the need to minimize spacecraft mass, size, power, and cost. Each satellite subsystem function must be examined to determine if reductions are possible, or if it can be combined with other functions or even eliminated. In addition, to generate such a fleet of nanosatellites on a normal development schedule requires that they be mass-produced, a process that would be facilitated by highly integrated instrumentation and spacecraft subsystems. Several of the technologies required for MagCon will be flight-tested by the New Millennium Program's Space Technology 5 mission, which is planned for launch in 2004 and will place three nanosatellites (mass = 19 kg) into geostationary orbit (see Figure 3.2).

To reduce the mass and power requirements of satellite subsystems means that some emphasis must be placed on new components and technologies, such as microelectromechanical systems (MEMS) and microthrusters (thrusters on a chip) for attitude control.[1] It also means developing much denser, more capable, and lower-power satellite data processing and control subsystems. Much effort is currently being put into developing low-power, high-density, high-speed application-specific integrated circuits, which cover the gamut of analog, digital, and hybrid analog/digital devices. It is hoped that they will achieve a significant reduction in satellite mass and power resource requirements as well as in the overall cost of the systems.

Other areas of technological concern include the injection of a large number of spacecraft in pre-determined orbits from one or a few "dispenser payloads"[2] and station keeping to maintain optimum spacecraft positioning.

Recommendation: NASA should continue to give high priority to the development and testing of advanced spacecraft technologies through such programs as the New Millennium Program and its advanced technology program.

ADVANCED SCIENCE INSTRUMENTATION

The development of highly miniaturized sensors of charged and neutral particles and photons will remain a critical and continuing focus of space-

TECHNOLOGY DEVELOPMENT

FIGURE 3.2 The New Millennium Program's Space Technology 5 (ST 5) mission, planned for launch in 2004, will test advanced spacecraft technologies required for future space physics missions involving constellations of miniaturized satellites. Lessons learned from ST 5 are expected to be incorporated in the Magnetospheric Constellation mission, a Solar Terrestrial Probe mission that will deploy 50 to 100 small satellites to study the structure and dynamics of Earth's magnetotail. Courtesy of NASA Goddard Space Flight Center.

research-related technology development during the coming decade. Advances in instrumentation generally provide increases in capability with a reduction in the required resources. Space science instrument development both drives advanced technology and utilizes emerging technologies at the forefront of materials science. Miniature particle multipliers and counters and charge-coupled device arrays for photon detection are two examples of emerging technologies that have significantly advanced space instrument capability. MEMS approaches can be used for adaptive apertures that extend the dynamic range of detectors. Miniaturization of micro-channel plate detectors and amplifiers on silicon substrates will allow low-power, high-sensitivity particle detectors to be constructed. Microcalorimeters (superconducting tunnel junctions and transition edge sensors) may be fabricated as pixel arrays, allowing unprecedented spatial and spectral resolution at wavelengths ranging from the infrared to the hard x ray. Doppler imaging of atmospheric emissions, allowing large-scale descriptions of charged and neutral dynamics, is currently being examined. Carbon annotates are being investigated to produce a low-power source of cold electrons for mass spectrometry and other applications.

Several programs within NASA's Office of Space Science fund instrument definition and development activities. The Planetary Instrument Definition and Development Program (PIDDP) supports studies of instrument concepts that will lead to the development of flight instruments to be proposed for specific planetary missions. While most of the instruments that have come out of this program are designed for the investigation of planetary atmospheres and surfaces, some space plasma physics instruments have been developed with PIDDP support as well, e.g., the energetic neutral atom component of the Magnetospheric Imaging Instrument on the Cassini Saturn Orbiter and the "hockey puck" energetic particle detector to be flown on MESSENGER. Such examples are few, however. Until recently, a program equivalent to the PIDDP did not exist for instrumentation for solar and space physics applications. This situation changed with the NRA 02-OSS-01, which provides for both a Sun-Earth Connection instrument development program and a Living With a Star geospace instrument development program.

Recommendation: NASA should continue to assign high priority, through its recently established new instrument development programs, to supporting the development of advanced instrumentation for solar and space physics missions and programs.

GATHERING AND ASSIMILATING DATA FROM MULTIPLE PLATFORMS

Missions involving constellations of spacecraft will place challenging requirements on operations and data acquisition systems. The need to resolve the spatial and temporal scales that describe the dominant physical processes in solar system plasmas has now become of paramount importance. The description of future flight missions inevitably includes the concept of multipoint measurements in the regions of interest. Such multipoint measurements, whether in the outer magnetosphere, the solar wind, or low-Earth orbit, offer unique challenges to spacecraft command and data-handling systems. Previous missions generally involved single autonomous spacecraft with data retrieval that was dependent only on scheduling a suitable contact time from a single ground station. Future missions will involve operation of many satellites that may be triggered by information from just one, and data transfer to and from the ground must accommodate multiple satellites in the same orbit with conflicting contact times. Such operational scenarios already exist, for example, in satellite telephone systems, but they have yet to be implemented in NASA science missions. The success of such future initiatives will require examination of the effectiveness of the Deep Space Network for many near-Earth missions. The construction of medium-range antennas might be more appropriate for such missions, and an appropriate distribution of more modestly sized antennas would allow some flexibility in the data acquisition plans. Finally, the growing number of receiving stations for missions in low Earth orbit will require coordination so that data from many sources can be consolidated and become transparent to the end user.

Recommendation: NASA should accelerate the development of command-and-control and data acquisition technologies for constellation missions.

MODELING THE SPACE ENVIRONMENT

The coming decade will see the availability of large space physics databases that will have to be integrated into physics-based numerical models. Data assimilation techniques were first used in numerical weather prediction. Meteorologists were confronted with having to solve an initial value problem without the correct initial data. Specifically, synoptic observations were insufficient to initiate a model run that could be used to predict the weather. To overcome this obstacle, meteorologists developed a methodology that is now called "data assimilation." This methodology uses data obtained at various places and times, in combination with a physics-based (numerical) forecast model, to provide an essentially time-continuous "movie" of the atmosphere in motion. During the last 40 years, meteorologists have continually improved their ability to predict the weather, as a result of both model improvements and a large infusion of new satellite and ground-based data. Following the example set by meteorologists, oceanographers began to use data assimilation techniques about 20 years ago, reaching the point where they were able to successfully predict the coming of the last El Niño.

Owing primarily to the relatively limited number of measurements of the near-Earth space environment, the solar and space physics community has not until recently had to address the issue of data assimilation as seriously as have the meteorologists. However, this situation is changing rapidly, particularly in the ionospheric arena. Within 10 years, it is anticipated that vast arrays of data sets will be available for assimilation into specification and forecast models such as those to be developed in the NASA Living With a Star theory and modeling initiative. The data include in situ and imaging data from numerous satellites, altitude profiles from a large network of ground-based ionosondes, and line-of-sight total electron content (TEC) and optical measurements between constellations of low-Earth-orbit and geostationary satellites and between the satellites and thousands of ground stations.

An increasing emphasis is being placed on understanding space science from a systems perspective, which ultimately requires access to reliable geophysical and solar-heliospheric parameters without regard to their origin or the instrumental details that underlie their derivation. These data will appear in a variety of formats and apply to quite disparate regions of space, and yet they must be accessible in a format that allows scientists to view and/or manipulate them in a manner dictated by the analysis procedures employed. The increased use of physics-based data assimilation will also require that renewed attention be given to data quality and to the generation of uncertainty estimates. The planned constellation-type missions, for example, will require parallel data assimilation techniques through which the spacecraft data can be directly injected into global magnetohydrodynamic models of the magnetosphere.

Recommendation: Existing NOAA and DOD facilities should be expanded to accommodate the large-scale integration of space- and ground-based data sets into physics-based models of the geospace environment.

OBSERVING GEOSPACE FROM EARTH

A global network of tended and autonomous ground-based ionospheric and geomagnetic measurement stations will continue to be an invaluable source of data on Earth's space environment during the next decade. Since the International Geophysical Year, ground-based observatory networks, particularly at high geographic latitudes, continue to be important. Several programs, such as the Canadian Auroral Network for the OPEN Program Unified Study (CANOPUS), have provided long-term observation facilities that are sometimes widely distributed in latitude, but nearly always localized in longitude.[3] Coordination provided by international programs such as the International Solar-Terrestrial Physics program has allowed for more widespread coverage by probes such as ionospheric radars and riometers,[4] magnetometers, all-sky cameras, and meridian-scanning photometers. In the near future it would be extremely beneficial for the United States to have such a network that is operating continuously and providing data in real time.

Tomography has been used extensively by the medical community for several decades, but it was not until about 1988 that this technique was first applied to the ionosphere. With radio tomography, transmissions from a low-Earth-orbiting satellite (or satellites, such as the planned Republic of China Satellite (ROCSAT) array) are received along a chain of stations with the intent to determine the TEC along the ray paths. Optical tomography works in a similar way, but uses airglow emissions instead of TECs. Currently, tomography chains in the United States, South America, Europe, Russia, and Asia provide information about ionospheric weather in these local regions. During the next decade, tomography is anticipated to play an important role in elucidating ionospheric weather features.

While the large radar facilities still require human tending, many of the smaller sensor systems and arrays have been deployed autonomously in the polar regions. However, the severe environments of temperature and moisture and the wildly varying solar insolation[5] have posed a serious reliability problem for these systems, to the point that their existence is now threatened.

Recommendation: The relevant program offices in the NSF should support comprehensive new approaches to the design and maintenance of ground-based, distributed instrument networks, with proper regard for the severe environments in which they must operate.

OBSERVING THE MAGNETIC SUN AT HIGH RESOLUTION

Developing the capabilities needed for the high-priority Advanced Technology Solar Telescope (ATST) requires that several important technical issues be addressed. Knowledge of the Sun has reached the point where a new, large-aperture solar telescope is needed to make observations of small-scale interactions between magnetic fields and plasmas.[6] Recent breakthroughs in adaptive optics have eliminated the major technical impediments to making observations with sufficient resolution to measure the pressure scale height and the photon mean free path length, the fundamental length scale in the solar atmosphere. The proposed ATST is a 4-m-class adaptive optics system that is designed to make the required measurements.

Development of the ATST presents several challenges not faced with large nighttime telescopes. The enormous flux of energy from the Sun makes thermal control a paramount consideration, both to remove the heat without degrading telescope performance and to control mirror seeing. To achieve diffraction-limited performance, a powerful adaptive optics system is required that operates at wavelengths from the visible to the infrared using solar structure as the wavefront sensing target. Low scattered light is essential for observing the corona and for measuring accurately the physical properties of small structures in sunspots, flux tubes, and magnetic pores.

While these challenges are well understood, a systematic program of technology development will be required in support of a timely development of the ATST. Progress in this direction is already under way as the National Solar Observatory (funded mostly by the NSF) has invested substantial resources in demonstrating the first solar adaptive optics system, which works with solar granulation as the wavefront sensing target.

Recommendation: The NSF should continue to fund the technology development program for the Advanced Technology Solar Telescope.

NOTES

1. Miniaturized ion thrusters are one of the technologies to be tested in the NMP's Space Technology 7 mission.

2. Multiple satellites (<10) have been dispersed from single launchers in the past (e.g., Iridium and Global Star); however, such projects have not had the spacecraft distribution requirements that are envisioned for MagCon. Significant effort is required to develop the dispenser ship technology by the end of the decade, when it will be required by MagCon.

3. CANOPUS is a Canadian geomagnetic and ionospheric observatory chain developed for NASA's Origins of Plasmas in Earth's Neighborhood (OPEN) mission, which was renamed the Global Geospace Science mission.

4. A riometer (relative ionospheric opacity meter) is a very sensitive radio used to study changes in the ionosphere by measuring the absorption of cosmic radio noise.

5. Insolation is defined as the direct or diffused short-wave solar radiation that is received in Earth's atmosphere or at its surface.

6. National Research Council, *Ground-Based Solar Research: An Assessment and Strategy for the Future*, National Academy Press, Washington, D.C., 1998.

4
Connections Between Solar and Space Physics and Other Disciplines

Solar physics and space physics have as their principal objects of study phenomena that occur in fully or partially ionized matter in the plasma state. These fields are related on a fundamental level to laboratory plasma physics, which directly investigates basic plasma physical processes, and to astrophysics, a discipline that relies heavily on understanding the physics unique to the plasma state. Solar physics and space physics share with laboratory plasma physics and astrophysics an interest in a variety of phenomena, including magnetic dynamo action, magnetic reconnection, turbulence, collisionless shocks, energetic particle transport and acceleration, and plasma instabilities. Understanding developed in one of these fields is thus in principle applicable to the others, and productive cross-fertilization between disciplines has occurred in a number of instances, for the reason that any fundamental principle, regardless of where it is discovered, is applicable throughout the universe.

At one time exploratory fields, solar and space physics and laboratory plasma physics have become mature disciplines probing fundamental scientific questions at levels of depth and sophistication that allow them to provide reality checks on one another and on astrophysics. Laboratory experiments have been developed to model particular processes occurring in space and astrophysical plasmas and to test the results of theoretical and computational studies. In situ and remote-sensing observations in the solar system have furnished a basis for the development of theoretical insights that in turn have been applied to astrophysical systems, while structures and processes observed in solar system plasmas have served as analogues for phenomena that may be occurring in astrophysical environments inaccessible to detailed study.

The various plasma regimes that are the focus of inquiry for solar and space physics, laboratory plasma physics, and astrophysics are characterized by a vast range of densities, temperatures, magnetic field strengths, and scale lengths. A key factor in applying insights from one discipline to another is the degree to which appropriate scaling relations can be established between the plasma regimes in question. Here, depending on the particular phenomena under investigation, commonly used dimensionless quantities or other ratios between relevant physical parameters may be of greater importance than dimensional quantities such as density or temperature, and meaningful scaling may link apparently quite different plasma environments, even ones that differ in scale by the orders of magnitude that separate solar system plasmas from astrophysical plasmas. For some astrophysical plasmas, such as pulsar magnetospheres or the degenerate plasmas of white dwarf interiors, temperatures or densities are so extreme that descriptions of the systems must include relativistic or quantum effects, and simple scaling from the regimes explored within the solar system may not be possible.

Although solar physics and space physics are concerned primarily with electromagnetic effects and matter in the plasma state, their relevance also extends into the domain of atmospheric science and climatology. Some of the points of contact with these fields are therefore also reviewed here. Finally, the committee considers briefly the role played by theoretical and laboratory studies of atomic and molecular processes in solar and space physics research.

LABORATORY PLASMA PHYSICS

Laboratory plasma experiments represent a valuable tool for advancing our understanding of the physical processes underlying phenomena observed in both solar system plasmas and remote astrophysical systems. The utility of such experiments is described in *Plasma Science*, a 1995 report by the National Research Council, which recommends establishment of an initiative to support laboratory studies relevant to space plasmas.[1] Benefits noted in the report include controllability, repeatability, precision, and the capacity for multipoint measurements and for observation of the temporal evolution of the process being studied. Laboratory plasma investigations are particularly important in that they provide a means of testing experimentally, under well-controlled conditions not achievable in space, the results of theoretical and computational studies. Such experiments can thus help to validate our theoretical understanding of space plasma processes. For example, the Magnetic Reconnection Experiment (MRX) at the Princeton Plasma Physics Laboratory has demonstrated good agreement between the reconnection rate observed in a laboratory plasma and that predicted by the

Sweet-Parker model of slow reconnection. The experiment has also demonstrated the occurrence of faster reconnection in circumstances reminiscent of the rapid reconnection observed in collisionless space plasmas.

To be sure, laboratory plasmas are critically affected by boundary effects and by initial conditions, and the densities and temperatures characteristic of space plasmas cannot be reproduced in the laboratory. For certain processes, however, the properties of a physically interesting space plasma can be modeled in a laboratory by selecting plasma parameters that approximate the dimensionless ratios that characterize the physics of the space plasma. Laboratory experiments have recently made particularly noteworthy contributions on two topics—magnetic reconnection and magnetic dynamo action.[2]

Magnetic Reconnection

Reconnection is being studied in several dedicated laboratory experiments around the world, including MRX and the Swarthmore Spheromak Experiment (SSX) (see sidebar, "Reconnection"). In each experiment, magnetized plasma loops are generated and merged. Magnetic energy is rapidly annihilated and converted to plasma flows, energetic particles, and heat. At present, the physical scale of such experiments is 0.1 to 1 m, and the key dimensionless parameter, the ratio of the time for the magnetic field to diffuse through a volume of plasma to the time for changes in field and plasma properties to propagate through the plasma, can be as large as ~1,000. Although this is much smaller than the corresponding ratio in magnetospheric plasmas, it is large enough to capture some of the qualitative aspects of the important physical processes. Diffusion is facilitated by particle collisions, and laboratory experiments can explore the transition from the collisional to the collisionless regime. Recent key results include the observation of basic two- and three-dimensional reconnection geometry, the observation of Alfvénic jets and super-Alfvénic particles, evidence that the thickness of the reconnection boundary layer is imposed by ion dynamics, and the demonstration of heating through the dissipation of magnetic energy in the reconnection layer.

Dynamo Experiments

In a dynamo, the motions of an electrically conducting fluid generate magnetic fields, thereby converting kinetic energy of motion into magnetic energy. Magnetic dynamos are critical elements of both solar and space

RECONNECTION

Magnetic reconnection is the fundamental mechanism by which magnetic energy is dissipated in the universe. Observationally, energy is released in bursts rather than in a continuous manner, driving phenomena such as solar flares and magnetospheric substorms.

The basic process of reconnection has been understood from the late 1950s. If two parcels of magnetized plasma have oppositely directed magnetic fields and there is a region of weaker or zero field between them, then under the right circumstances the parcels can approach each other. The oppositely directed magnetic flux can cancel out (annihilate), and the plasma can jet outward along the weaker field directions at a characteristic speed called the Alfvén speed.

Magnetic flux is lost from the structures on the "inflow" sides, while a new magnetic structure, formed from "reconnected" magnetic field lines, grows on the "outflow" sides. As the plasma approaches the central region, the magnetic field may change direction very quickly, producing intense channels of electric current density that can heat the plasma. The current typically takes the form of filaments or sheets, and the magnetic field often takes on a characteristic X-point shape (as illustrated on the facing page). This process is called magnetic reconnection or merging, and it is thought to be the main means by which magnetic fields in space plasmas change the way they are linked with one another.

Recent observations in the magnetosphere and in the solar corona provide mounting evidence of the key role of reconnection in space plasmas. There are important outstanding questions concerning the small-scale structure of the reconnection region, in which plasma kinetic effects are dominant. In these crucial small-scale regions, particles can be accelerated to high energies, magnetic field lines break and reconnect, plasma jets are formed, heat is released, and energy can be transferred from one region to another. In reconnection, large-scale dynamics and small-scale plasma physics come face to face: This is an essential feature of multiscale, nonlinear space plasma physics.

---

plasmas, but there is no fully satisfactory theoretical understanding of solar and stellar dynamos.[3] Several laboratory experiments around the world are investigating the dynamo problem, and the results of such investigations afford insight into the process. In each, the kinetic energy of an electrically conducting fluid is converted (in part) to large-scale magnetic field energy. Typically, the working fluid is sodium (a fluid at 100°C). Currently, the physical scale of such experiments is 0.1 to 1 m, and the key dimensionless parameter, which in this case is the ratio of the time for the field to diffuse

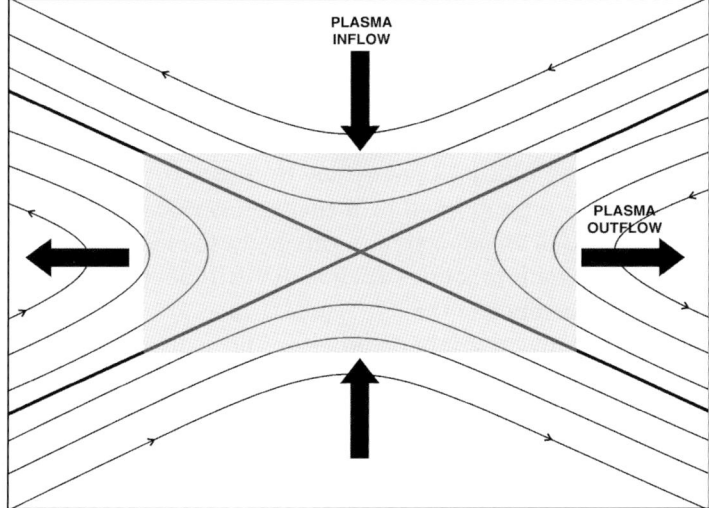

Schematic illustrating the reconnection process. Oppositely directed magnetic fields are "frozen" into the plasma flow and carried toward the reconnection site. Here, in a localized region of space known as the diffusion region (gray box), the magnetic flux and the plasma become decoupled, allowing the magnetic field lines to break and reconnect. The kinetic processes that occur in the diffusion region to facilitate reconnection are the subject of ongoing theoretical and modeling studies and will be studied in the geospace environment by the Magnetospheric Multiscale mission.

through a volume of plasma to the time for the fluid to flow across it, is about 10 to 50. Here, as in the reconnection studies, the laboratory plasma is more dissipative than are most typical space plasmas, but the laboratory experiments demonstrate and explore the physical effects that are relevant to the physical systems of interest to solar and space physics. Progress has been made in identifying the natural oscillations of the system, referred to as eigenmodes, observing the decay of excited eigenmodes, and measuring the effects of turbulence on electrical conductivity.

Other Problems

Other plasma processes important to space physics such as the generation and propagation of waves unique to magnetized plasmas (e.g., whistler and shear Alfvén waves) have been investigated in laboratory experiments. In one experiment using the Large Plasma Research Device of the University of California at Los Angeles, shear Alfvén waves were generated within field-aligned density depletions under plasma conditions comparable, in terms of the relevant parameters, to those in the ionosphere. Interesting changes in wave properties have been observed as the waves propagate into regions of changing plasma properties, and these experiments are helping to elucidate processes such as particle heating and acceleration in the auroral ionosphere.

Global magnetospheric processes have also been studied in the laboratory. In particular, terrella experiments have been carried out at the University of California, Riverside to investigate the effects of the solar wind/magnetosphere interaction on the distant magnetotail. Although the scaling between the laboratory magnetosphere and Earth's magnetosphere is not as consistent as desired, the terrella experiments have provided qualitative support for the predicted development of X-type (Y-type) neutral lines under conditions of a southward (northward) interplanetary magnetic field and for the occurrence of high-latitude reconnection when the magnetic field in the incident plasma (representing the solar wind) points northward. Other plasma physical phenomena that have been the subjects of recent laboratory experiments include waves in dusty plasmas and solar prominences.

Recommendation: In collaboration with other interested agencies, the NSF and NASA should take the lead in initiating a program in laboratory plasma science that can provide new understanding of fundamental processes important to solar and space physics.

As noted above, the establishment of such a laboratory initiative was previously recommended in the 1995 National Research Council report *Plasma Science*.

ASTROPHYSICAL PLASMAS

The solar system is home to a rich variety of plasmas, ranging in density from the collisional plasma ($\sim 10^{26}$ cm^{-3}) of the Sun's interior to the attenuated, collisionless medium of the outer heliosphere (~ 0.002 cm^{-3} at 50 AU) and in energy from the cold $\sim$1-eV plasma of Earth's plasmasphere to the

MeV trapped particles of Jupiter's synchrotron-emitting radiation belts. In addition to the highly ionized plasmas of the interplanetary medium and planetary magnetospheres, the inventory of solar system plasmas includes partially ionized gases, as in the terrestrial ionosphere, where ion-neutral collisions strongly influence the electrodynamics, and dusty plasmas in comet dust tails and in planetary ring systems, where the effects of gravity and/or radiation pressure must be considered as well as electromagnetic influences. In contrast to remote astrophysical plasma systems, solar system plasmas are accessible both to direct, in situ measurement and to systematic, highly resolved remote sensing, and, in contrast to laboratory plasmas, boundaries can be remote in solar and space plasmas. Further, the physical properties of solar system plasmas span an enormous range, with analogies in numerous other astrophysical environments. The solar system thus represents a laboratory for astrophysical studies in which fundamental plasma physical processes such as reconnection, turbulence, particle acceleration, and shocks can be investigated at both the micro- and macroscales and in ways not possible through numerical simulations or laboratory plasma experiments.[4] The physical understanding derived from observational studies of solar system plasmas provides a basis for theoretical extrapolation to more extreme astrophysical systems. Fruitful cross-fertilization between solar-system plasma physics and astrophysics is particularly well exemplified in the study of collisionless shocks, magnetohydrodynamic turbulence, and magnetic reconnection.

Collisionless Shocks

The Panel on Opportunities in Plasma Science reported in 1995 that the "study of collisionless shocks is arguably the area in which the most significant advances have been achieved during the past decade and where the impact of space on basic plasma physics has been most profound."[5] Substantial advances in this area were based on in situ observations of Earth's bow shock, of bow shocks at other planets and comets, and of interplanetary shocks. The observational and associated theoretical studies of shocks in the solar system established critical properties relevant to collisionless shocks in the interstellar medium and have illuminated the mechanisms through which such shocks produce particle acceleration. Diffusive shock acceleration, in which charged particles gain energy through repeated cycles of scattering and shock crossing, is a robust mechanism now thought to be responsible for energetic particles throughout the universe. Some astrophysical shocks such as those driven by young supernova remnants have Mach numbers greater than 100. They are stronger than the Uranian bow shock, which, with a magnetosonic Mach number of 17, is the strongest shock encountered in the solar system. The character of a colli-sionless shock can change with increasing Mach number, and in extrapolating beyond the range of observations, phenomena not present in weaker shocks (turbulence, for example) may modify the form of shock-controlled particle acceleration. In seeking to understand astrophysical shocks stronger than those encountered in the solar system, the results of analytical theory and numerical simulations will be of critical value.

Magnetohydrodynamic Turbulence

Magnetohydrodynamic (MHD) turbulence is characterized by nonlinear interactions among fluctuations of the magnetic field and flow velocity over a range of spatial and temporal scales. It plays an important role in plasma heating, the transport of energetic particles, and radiative transfer and is ubiquitous in space and astrophysical plasmas. The solar wind and the diffuse interstellar medium (ISM) are both examples of plasmas that exhibit turbulent behavior, as evidenced by the power spectra determined from radio propagation observations (ISM, solar wind) and in situ data (solar wind) (see sidebar, "Turbulence in the Solar Wind"). These observations have stimulated an ongoing effort to develop theoretical treatments of MHD turbulence appropriate to the two similar systems. Although this effort relies heavily on studies of the solar wind as the more fully characterized of the two plasmas, it is strongly interdisciplinary in character, drawing on insights from theoretical plasma physics and laboratory plasma physics as well as on observational and theoretical studies. The last 20 years have seen considerable progress in this area. However, a number of important theoretical issues remain to be resolved—for example, regarding scaling in MHD turbulence and compressibility effects.[6] In addition, the mechanisms responsible for ISM turbulence are poorly understood, and heliospheric analogies are expected to provide useful insights. Supersonic turbulence is studied in the star-forming regions in dense molecular clouds, and it is noteworthy that turbulent energy decay rates are found to scale much the same way as in the heliosphere, suggesting a commonality of MHD turbulence principles. Random electric fields associated with turbulence can also play a role in the stochastic acceleration of charged particles, and this is a possible mechanism for reacceleration of cosmic rays in the Galaxy.

Magnetic Reconnection Theory

Reconnection theory is an especially compelling example of a profound contribution made by solar and space physics to astrophysics, as well as to basic plasma physics and fusion research. The concept of magnetic reconnection was initially developed principally in the context of efforts to explain the production of solar flares. Reconnection is now considered to offer a plausible explanation for a number of other solar phenomena as well, such as transition-region explosive events, surges and sprays, and x-ray bright points. Since the early 1960s, reconnection theory has been extensively and successfully applied, in the open model of the magnetosphere, to the related problems of energy, mass, and momentum transfer from the solar wind into the magnetosphere and the explosive release of stored magnetic energy in substorms. Recent theoretical studies and numerical simulations have yielded new insights into the microscopic aspects of the reconnection process. Finally, theoretical reconnection models are coming into correspondence with observations and experimental results on the subject, promising rapid progress in the coming decade.

The concept of reconnection has been invoked to address numerous problems in astrophysics. One of its earliest applications was to the problem of magnetic flux reduction in gravitationally collapsing protostellar clouds as part of the star formation process. Flux removal by reconnection has also been proposed as relevant to the production of magnetic viscosity in accretion disks. Magnetic reconnection is thought to be responsible for flarelike releases of energy in interactions between neutron star magnetospheres and accretion disk magnetic fields as well as for flares occurring in accretion disk coronae, and the solar flare model of reconnection has been applied to flares from dwarf stars, binaries, and T Tauri stars. Recent studies suggest that reconnection may plausibly play a role in rapid particle acceleration—e.g., in the reacceleration of high-energy electrons in extragalactic jets and in the acceleration of ultrahigh-energy cosmic rays. Such a mechanism would be similar to direct acceleration of charged particles by solar flares. Fast reconnection in a magnetized relativistic outflow has also been put forward as a candidate mechanism for the production of radiation in gamma-ray-burst fireballs. Numerous other examples of reconnection in astrophysical phenomena can be found in the recent literature, documenting the importance of this fundamental concept. Applications of reconnection theory—as well as of the knowledge obtained through the study of turbulence and shocks in the solar wind—to astrophysical phenomena exemplify the increasingly wide recognition within the astrophysics

TURBULENCE IN THE SOLAR WIND

Spacecraft venturing into the solar wind measure broadband fluctuations in all plasma variables, including the proton fluid velocity, the density, and the magnetic field. Such fluctuations are in many ways reminiscent of ordinary fluid turbulence; however, turbulence in the solar wind is also distinctly a plasma phenomenon, involving magnetic fields, kinetic effects, and interactions with charged particles. The observed turbulence is thought to enhance heating of the interplanetary medium, and it is responsible for scattering and transport of cosmic rays originating from both inside and outside the heliosphere.

Solar wind turbulence also provides important clues about the nature of the lower solar corona. Fluctuations are observed over a wide range of spatial scales, from the system size (AUs in size) down to electron kinetic scales, 100,000 times smaller. Large-scale fluctuations can be described by fluid models, but the small-scale activity (less than the proton kinetic scales) requires a kinetic description.

The physics of the couplings between large-scale and kinetic-scale processes in the solar wind is mediated by properties of the turbulence. A substantial portion of the turbulence energy is distributed in a distinctive power-law spectral distribution over about three decades of spatial scale, and is highly reminiscent of ordinary fluid turbulence. Interacting waves originate in the corona; and in interplanetary space, turbulence is further enhanced by instability and interactions with shear layers near high-speed solar wind streams. So begins a cascade in which turbulent fluctuations, including eddies and waves, interact with one another to produce smaller-scale fluctuations. Eventually these are damped by kinetic processes and heat the plasma. With an explicitly turbulence-based model, the radial evolution of turbulence intensity and the temperature of the plasma can be computed.

Dissipative turbulence is driven by shear in solar wind streams and by pickup ion effects. Such a model can account for observed amplitudes and temperatures from 1 AU to beyond 60 AU (see figure on the facing page). Like its terrestrial counterparts, solar wind turbulence remains an incompletely understood topic but an important one, in that it mediates complex dynamical couplings between very large and very small scales, very slow fluid motions and very fast kinetic processes, and very-low-energy and very-high-energy plasma particles.

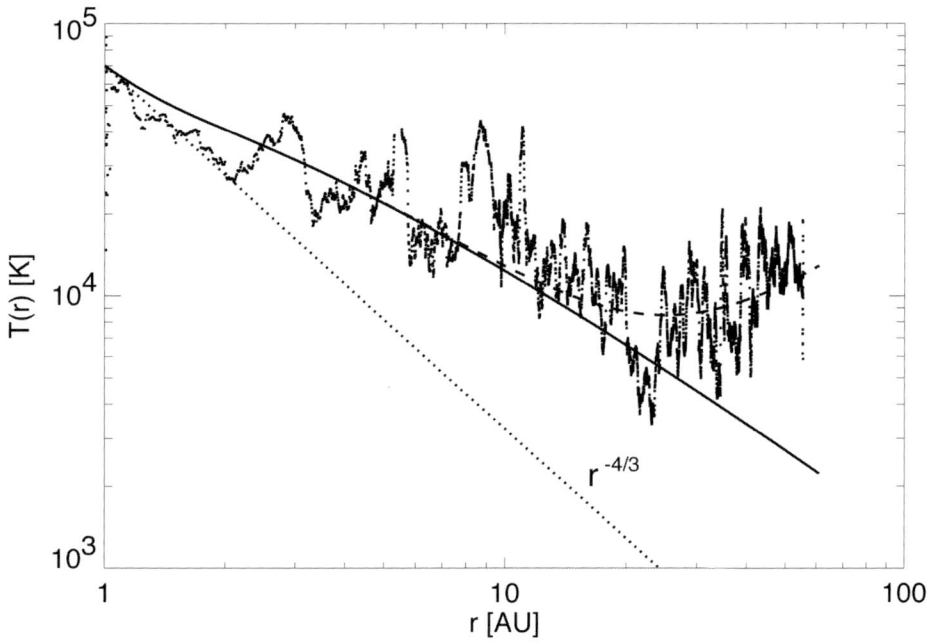

Proton temperatures from 1 AU to 60 AU as measured by Voyager 2. Temperatures are significantly hotter than those indicated by the adiabatic law (dotted) and can be described by a turbulence model (solid). By including pickup ions, the turbulence model captures the temperature increase from 20 AU to 60 AU, showing highly non-adiabatic behavior. Courtesy of W.H. Matthaeus. (This figure originally appeared in W.H. Matthaeus, G.P. Zank, C.W. Smith, and S. Oughton, *Physical Review Letters* 82: 3444-3447, 1999. Copyright 1999 by the American Physical Society.)

community of the importance of magnetic fields and electrodynamics in the structure and dynamics of astrophysical plasmas.

Solar and Stellar Processes

As the only well-resolved star, the Sun plays a uniquely important role as a laboratory for stellar processes. Helioseismology has confirmed the theoretical model of the solar interior and revealed complex, three-dimensional flow patterns within the convective zone, presumably responsible for generating and sustaining the Sun's magnetic fields. Precision confirmation of the theoretical model of the Sun gives us confidence in the theoretical models of the interior structure, dynamics, and evolution of other stars (see sidebar, "The Solar Laboratory"). Similarly, recent high-resolution observations of the Sun's highly structured and dynamic corona, with its magnetic loops, arcades, filaments, and fibrils, contribute to our understanding of other stellar x-ray coronas, of which the Sun is the prototype.

Despite considerable progress in our knowledge of solar processes, many fundamental questions of importance to solar physics and astrophysics have yet to be answered. For example, the detailed workings of the Sun's magnetic dynamo remain a major mystery—one whose eventual solution will have profound implications for our understanding of MHD dynamos not just in other solar-type stars but in other astrophysical settings as well. Still an open question some 40 years after the theory of the supersonic solar wind was formulated and confirmed by in situ measurement is how the corona is heated and the solar wind accelerated. In situ measurement of the near-Sun environment (inside 0.3 AU), combined with high-resolution remote sensing, promises to resolve this question and to yield insights into coronal processes at other late-main-sequence stars.

Finally, the committee notes that the cross-fertilization between solar physics and astrophysics is bidirectional. Studies of the activity and luminosity of other solar-type stars can validate our theories of the solar dynamo, provide a glimpse into the Sun's past and a preview of its future, and contribute to our efforts to understand secular variations in solar activity such as the Maunder Minimum and Medieval Maximum.

ATMOSPHERIC SCIENCE AND CLIMATOLOGY

All solar-system-body atmospheres are exposed to and interact with either the solar wind (Venus, Mars, Pluto) or the plasma contained within a magnetosphere (Earth, the giant planets, the Galilean moons, and the moons

THE SOLAR LABORATORY

Helioseismic data from SOHO/MDI and GONG are fast becoming a scientific treasure house that connects several fields. In the 1980s, there were detectable discrepancies between the Sun's observed seismic frequencies and those calculated from theoretical models of the Sun. With time, the measured seismic frequencies have become much more precise.

The frequency discrepancies led atomic physicists to improve the treatment of atomic physics of solarlike plasmas—opacities and the equation of state. This development, coupled with improved treatments of diffusion, has improved the theoretical model of the solar interior to the point that it predicts a sound speed that nowhere differs from that determined by helioseismology by more than one part in a thousand (the present uncertainty in the helioseimic speed of sound). The precision is now such that we can use the helio-seismic frequencies to test individual nuclear cross sections of the p-p chain, which are the source of the Sun's energy. These reactions are nearly impossible to measure in the laboratory because they are so low in energy.

For years, it has been known that the models of the Sun predict far fewer solar neutrinos than are observed. With the agreement between the theoretical solar model and the precise results from helioseismology, it is now clear that there is no apparent astrophysical solution to the Sun's neutrino deficit. Particle physicists have recently determined that the origin of the deficit is that standard electroweak physics is wrong—that is, neutrinos have mass.

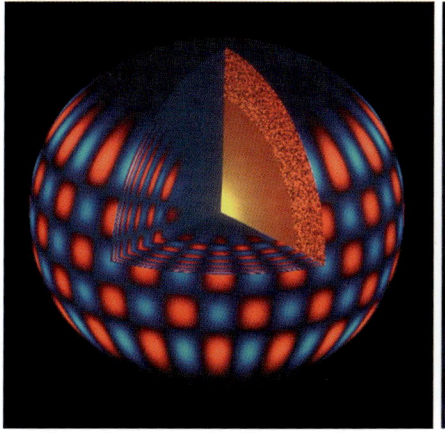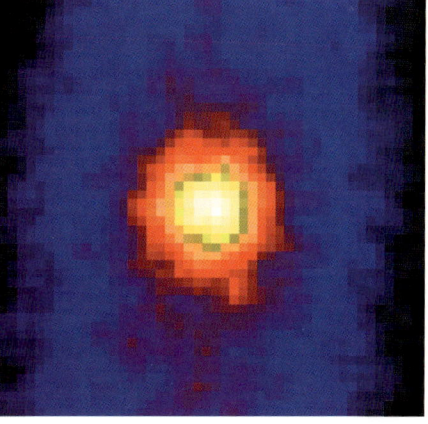

Left: An artist's concept of the solar oscillations used in helioseismology to probe the solar interior. Right: The neutrino flux from the Sun, observed with the Super-Kamiokande neutrino detector and integrated over a period of 500 days. The cutaway view of the Sun courtesy of J.W. Harvey (GONG Project, National Solar Observatory/AURA/NSF). The neutrino image courtesy of R. Svoboda (University of Louisiana).

Titan and Triton). The atmospheres are also exposed to bombardment by galactic cosmic rays. Thus there are numerous points of contact between space physics and atmospheric science, particularly in the area of aeronomy, which represents the interface between the two disciplines and which can arguably be considered a subdiscipline of either field. It is not an exaggeration to say that the structure and evolution of a planetary atmosphere cannot be understood without considering the effects of solar system plasmas.

The interaction of a planet's or satellite's atmosphere with the plasma environment affects the upper atmosphere, with its ionized and neutral components, most strongly. At magnetized planets this interaction occurs primarily between the ionosphere and the magnetosphere, which are electrodynamically coupled by field-aligned currents. The deposition of electromagnetic and particle kinetic energy and momentum by auroral processes modifies atmospheric structure, circulation, and composition. Auroral energy input dominates the energetics of Jupiter's high-latitude upper atmosphere and, during times of intense geomagnetic disturbance, the energetics of Earth's upper atmosphere as well.

In the case of unmagnetized or weakly magnetized bodies such as Venus, Mars, and Titan (the moon with the most massive atmosphere), auroral processes play no role in upper atmosphere aeronomy. The interaction of the ionospheres of such bodies with the ambient plasma does, however, have a significant atmospheric effect owing to the removal of ionospheric ions that are picked up by the external plasma and swept away in its flow. Sputtering by returning pickup ions is another important loss process driven by the atmosphere-plasma interaction at unmagnetized bodies. Theoretical studies have shown that the loss of atmospheric material as a result of scavenging and sputtering may have been substantial over the age of the solar system. Moreover, sputtering may also have contributed to the loss of the primordial atmospheres of Mercury and the Moon.

Sputtering not only plays a role in atmospheric loss but also, through the sputtering of surfaces, produces tenuous atmospheres such as the recently discovered oxygen exospheres about the jovian moons Europa and Ganymede. Sputtering of the surface of Mercury by solar wind particles probably contributes, along with other processes such as photodesorption, to the creation of Mercury's variable sodium exosphere. (Charged-particle bombardment can also affect the composition and properties of the impacted surfaces, a process known as space weathering. Investigation of this process is thus an important point of contact between space physics and solid-body studies as well as between space physics and the study of planetary atmospheres.)

The space environment influences the middle and lower atmospheres of some solar system bodies as well as their upper atmospheres, although in these regions the effects are generally less extensive than at higher altitudes. For example, bombardment by energetic electrons from Saturn's magnetosphere is a possible mechanism for producing hydrocarbon aerosols in Titan's haze layers. Similarly, energetic auroral particle precipitation is thought to lead to the formation of the high-latitude stratospheric hazes observed at Jupiter and Saturn. Modeling studies indicate that heating by such hazes may play an important role in the large-scale meridional circulation of Jupiter's stratosphere. Nucleation induced by ions created by galactic cosmic ray (GCR) bombardment has been shown to be a viable mechanism for the formation of hydrocarbon hazes and clouds in Neptune's lower stratosphere and troposphere; if GCR-induced ionization does indeed play a role in aerosol production at Neptune, Neptune's haze, and hence albedo, may vary in phase with the solar-cycle modulation of the cosmic-ray flux. In the case of Earth, the effect of solar energetic particle events on the ozone chemistry of the middle atmosphere is well known. Odd-hydrogen and odd-nitrogen production initiated by solar proton bombardment leads to the destruction of ozone in the polar mesosphere and upper (and sometimes middle) stratosphere. Most of the solar energetic particle-induced ozone loss is relatively short-lived; however, particularly intense events, such as those that occurred in August 1972, October 1989, and July 2000, can have longer-lasting (months to years) effects.

Other possible influences of the space environment on Earth's lower atmosphere should be noted. Ionization produced by galactic cosmic rays could have an influence on the nucleation of cloud particles. An apparent correlation between globally averaged low cloud cover and the cosmic ray flux over solar cycle 22 has been adduced in support of this mechanism. Another postulated influence involves solar-wind-modulated variations in the flow of current density in the global electric circuit, variations that cause changes in cloud microphysics. Both mechanisms provide a means of coupling solar/solar wind activity to weather and climate. However, while observations have occasionally been found to support these mechanisms, they remain ill substantiated and controversial. If they are real, solar-induced changes in cloudiness are likely to produce regional rather than global changes, in contrast to the effects of small variations in irradiance. However, the societal impacts of regional climate changes are important and justify continuing attention to understanding how, or indeed whether, the solar-cycle effects act as proposed.

In addition to such possible influences from space on Earth's weather and climate, the influence of global climate change on the geospace environment—at least on its lower reaches—must also be considered. Studies using the thermosphere-ionosphere general circulation model developed at the National Center for Atmospheric Research have demonstrated that the ionosphere-thermosphere system may experience long-term changes resulting from the buildup of anthropogenic trace gases in the middle atmosphere. By analogy with the term "space weather," these secular changes could appropriately be described collectively as "space climate." The predicted effect is principally one of mesospheric and thermospheric cooling due to increased concentrations of carbon dioxide and methane. In a reasonable scenario of trace gas increases, the global average temperature could decrease during the 21st century by ~10 K in the mesosphere and by ~50 K in the thermosphere. The consequences of such a change in the thermal structure of these regions would include changes in the densities of the major and minor neutral species, in ionospheric plasma density and temperature, and in the dynamics of the upper atmosphere.

ATOMIC AND MOLECULAR PHYSICS AND CHEMISTRY

Knowledge of the properties of atoms and molecules, ranging from their spectra to their cross sections for excitation, ionization, and charge exchange, is critical for our understanding of a number of magnetospheric, ionospheric, solar, and heliospheric processes. Such data are needed, for example, to understand how energetic magnetospheric particles interact with satellites such as the Galilean moons of Jupiter to produce their tenuous atmospheres and the heavy ion plasmas that characterize the giant magnetospheres. The interpretation of observations of auroral and dayglow emissions from the outer planets requires accurate excitation and emission cross sections of molecular hydrogen. Accurate modeling of complex processes such as EUV and collisional ionization, recombination, charge exchange, and electron-stripping of high-energy particles is essential for interpreting the ionic charge state and elemental composition of the solar wind. The observational determination of velocities in the corona requires precise wavelengths for the UV and EUV emission lines, and the determination of abundances of the diverse ions requires precise knowledge of the line strengths.

As these examples demonstrate, laboratory and theoretical studies of atomic and molecular properties are central to our understanding of important aspects of solar system plasmas. In many cases, however, our current knowledge of the atomic and molecular processes for a particular problem is inadequate, and further studies are required. For instance, the recently discovered x-ray emission from comets has been attributed to charge-transfer collisions between highly charged solar wind ions and cometary neutrals. However, the cross section information needed to understand this phenomenon is currently inadequate for the relevant collision species and at solar wind energies. Fortunately, a number of new laboratory investigations have recently been initiated in this area.

Recommendation: The NSF and NASA should take the lead and other interested agencies should collaborate in supporting, via the proposal and funding processes, increased interactions between reseachers in solar and space physics and those in allied fields such as atomic and molecular physics, laboratory fusion physics, atmospheric science, and astrophysics.

NOTES

1. National Research Council, *Plasma Science: From Fundamental Research to Technological Applications*, National Academy Press, Washington, D.C., 1995, pp. 25 and 119.

2. National Research Council, *An Assessment of the Department of Energy's Office of Fusion Energy Sciences Program*, National Academy Press, Washington, D.C., 2001, p. 67.

3. How Earth's magnetic field is generated has been reasonably well established by recent numerical experiments on the hydrodynamic convection of the liquid iron core of Earth. The only aspect of the problem that might need further study seems to be the heat source that drives the convection, currently believed to be the crystallization of the liquid iron at the surface of the central iron core. An obvious obstacle to developing an equally precise model for other planets is their relatively less well known interior structure.

4. This point has been made in previous NRC studies. The Colgate report (NRC, *Space Plasma Physics: The Study of Solar-System Plasmas: Volume 1, Reports of the Study Committee and Advocacy Panels*, National Academy Press, Washington, D.C., 1978, p. 9) states, for example,

> [In contrast to laboratory plasmas,] the plasmas in the solar system are examples of semi-infinite astrophysical plasmas. The boundary conditions are more relevant to astrophysics, and the parameter ranges are closer to those occurring in astrophysics. For this reason, solar-system plasmas provide a relatively convenient, accessible "laboratory" where many of the phenomena seen in astrophysics can be closely observed. The impact of this science on astrophysics has been significant and should be more so in the future.

5. National Research Council, *Plasma Science: From Fundamental Research to Technological Applications*, National Academy Press, Washington, D.C., 1995, p. 171.

6. In its 1995 report, the Panel on Opportunities in Plasma Science identified MHD turbulence as one of the outstanding problems in plasma astrophysics (NRC, *Plasma Science*, 1995, p. 124):

> Yet, we do not have a complete theory of hydromagnetic turbulence, and simulations, which are of great educational value, do not yet resolve the full range of relevant scales. Progress in understanding MHD turbulence will probably be made through a combination of direct observation (such as in situ measurements in the solar wind), simulations, and analytic theory.

5
Effects of the Solar and Space Environment on Technology and Society

Solar activity and Earth's space environment can have deleterious effects on numerous technologies that are used by modern society. Understanding the origin of these effects is essential for the successful design, implementation, and operation of modern technologies. Not only will future research in solar and space physics address high-priority, frontier science objectives, but it will also provide key data and new understanding essential for protecting vulnerable systems against the harmful effects of the space environment.

CHALLENGES POSED BY EARTH'S SPACE ENVIRONMENT

While a number of schemes for long-distance communications had been proposed and even developed to some extent before the implementation of a practical electrical telegraph, it was the working model by Samuel F.B. Morse in 1835 that led to the first revolution in communications. Telegraph systems were installed rapidly in Europe and in the eastern United States after that, and the first successful transatlantic cable was laid in 1866. Shortly after the installation of several telegraph lines there were indications that the systems appeared to be detrimentally affected by external factors. Intervals of "spontaneous" electrical currents were often measured on the lines but were not understood. Less than a day after Richard Carrington first observed a white-light solar flare in 1859, commercial telegraph systems in New England and Europe experienced severe outages and even some sporadic operations without the benefit of their battery supplies. Telegraph systems used for military communications in the American Civil War suffered disruptions at times. Such events suggested that Earth's environment, and perhaps even the Sun, caused these disturbances to telegraph systems. However, many scientists and telegraph engineers remained unconvinced for nearly a half century.

With the development in the late 19th and early 20th centuries of additional electricity-based technologies, including the telephone, it became increasingly apparent that the Sun was indeed the ultimate source of many telegraph and telephone system disturbances as well as of disturbances and outages in new technologies such as wireless communications and electrical power systems on local and larger geographical scales. Ironically, after the discovery of the ionosphere by Gregory Breit and Merle Tuve and by Edward Appleton, it also became evident that the Sun was the source of the ionization layer that facilitated early long-distance wireless transmissions, which had become especially important following Gulielmo Marconi's demonstration of transatlantic transmissions in 1901.

When Arthur Clarke and then John Pierce suggested the use of Earth-orbiting satellites for communications, they did not anticipate that the environment of space would be anything but benign. However, the discovery of Earth's charged-particle radiation belts by Van Allen in 1958 showed that this was not the case.

Concurrent with progress in spaceflight and in understanding of the space environment over the last four decades, increasingly sophisticated technologies have been developed and deployed on Earth and in space. What has been learned, often by hard experience, is that the space environment around Earth cannot be ignored in the design and operation of a wide range of systems and their components. Illustrated schematically in Figure 5.1 and listed in Table 5.1 are many of the processes occurring in space that can affect a large variety of ground- and space-based technologies.[1,2]

The harsh radiation environment of space is well recognized today, and all spacecraft—whether sent into orbit around Earth or to the far reaches of the solar system—must be designed to withstand the radiation environments that they may encounter. Anomalies produced by space radiation affect the operation of computer memory and processors aboard these spacecraft, and such anomalies seem to become more prevalent as semiconductor devices continue to shrink in size. Space radiation can also produce differential charging across spacecraft surfaces and in dielectric materials deep within the spacecraft itself, including cabling. If the charge buildup becomes large enough, electrical breakdowns (analogous to those caused by lightning discharges) occur that can damage sensitive electronic components and/or produce erratic signals in control lines.

In addition to the threat it poses to spacecraft systems, the radiation environment of space also poses risks for astronaut health and safety. Appropriate measures must thus be pursued to minimize the exposure of astronauts to particle radiation during geomagnetic storms and solar energetic

EFFECTS ON TECHNOLOGY AND SOCIETY 113

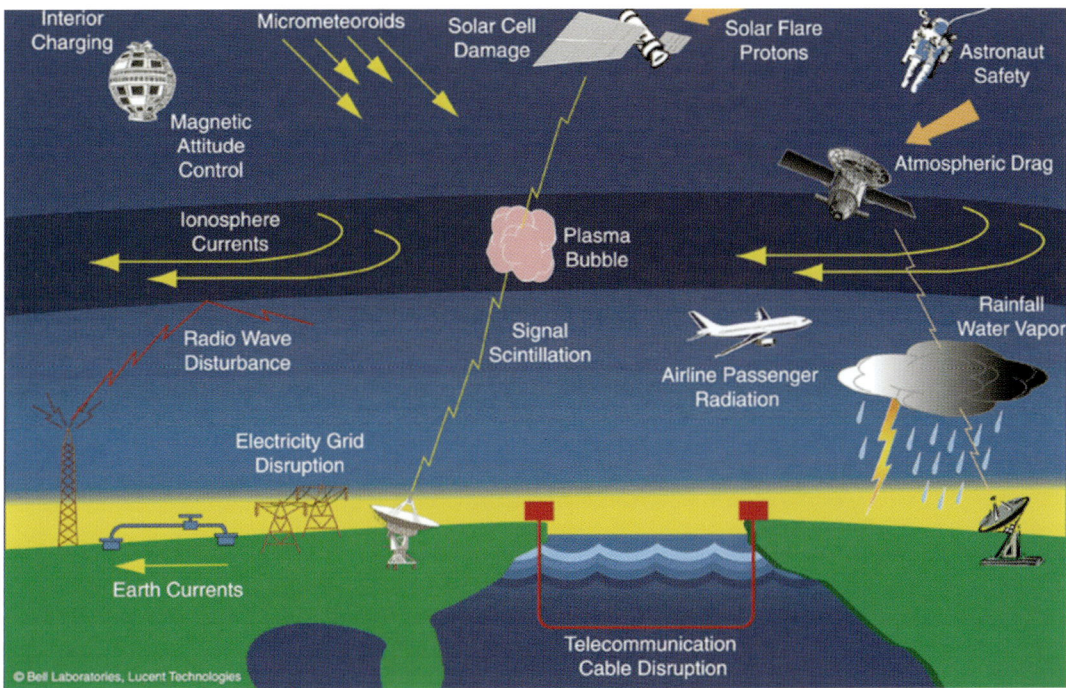

FIGURE 5.1 Triggered by solar activity, disturbances in Earth's space environment can adversely affect a variety of technological systems and present a hazard to the health and safety of astronauts as well. Understanding the physics of space weather and developing the means to predict its occurrence and to mitigate its effects are important national goals toward which such initiatives as NASA's Living With a Star program and the multiagency National Space Weather Program are directed. Courtesy of L.J. Lanzerotti (Lucent Technologies). (This figure was previously published in L.J. Lanzerotti, Space weather effects on technologies, in *Space Weather*, P. Song, H.J. Singer, and G.L. Siscoe, eds., Geophysical Monograph 125, 11-22, American Geophysical Union, Washington, D.C., 2001. Reproduced by permission of the American Geophysical Union.)

---

particle events. Solar energetic particle events could also be hazardous to the health of airline crews and passengers on polar routes.

While ionospheric disturbances have long been known to both facilitate and impair high-frequency wireless transmission and reception, such disturbances have also been found to interfere with transmissions to and from orbiting spacecraft using frequencies near a gigahertz and even higher. Such effects (generally "scintillation" of the signals—an effect that can be very disruptive) are found to be most severe in the equatorial and polar regions. Scintillation and other ionospheric disturbances are of serious

TABLE 5.1 Impacts of Solar-Terrestrial Processes on Technology

Ionospheric variations
    Induction of electrical currents in Earth
        Malfunctioning of power distribution systems
        Malfunctioning of long communications cables
        Malfunctioning of pipelines
    Interference with high-frequency communications
    Interference with the Federal Aviation Administration's wide area augmentation system
    Interference with surveillance, over-the-horizon radars, and radar altimetry
    Interference with geophysical prospecting
    Enabler of geophysical prospecting
    Interference with communications satellite signals scintillation
Magnetic field variations
    Interference with attitude control of spacecraft
    Interference with compasses
Solar radio bursts
    Excess noise in wireless communications systems
Radiation
    Solar cell damage
    Semiconductor device damage and failure
    Faulty semiconductor devices
    Spacecraft charging: degradation of surface and interior materials; production of electrical noise and disturbances
    Risks to astronaut safety
    Risks to airline crew and passenger safety
Micrometeoroids and space debris
    Solar cell damage
    Damage to mirrors, surfaces, materials, complete vehicles
Disturbances of the upper atmosphere
    Interference with low-altitude satellite tracking and lifetimes
    Attenuation and scatter of wireless signals

SOURCE: Adapted from L.J. Lanzerotti, Space weather effects on technologies, in *Space Weather*, Geophysical Monograph 125, P. Song, H.J. Singer, and G.L. Siscoe, eds., American Geophysical Union, Washington, D.C., 2001, pp. 11-22.

concern for military and civilian communications and navigation (e.g., for the use of single-frequency global positioning devices in applications that demand high precision.)

Although the telegraph with its extended network of lines and cables is an obsolete technology and the effects of the space environment on it are now irrelevant, solar-activity-induced disturbances still produce detrimental changes across other long conductors anchored to Earth, including pipelines, long-haul telecommunications lines, and glass-fiber systems that need electrical power for signal regeneration. Electrical power grids continue to become more complex and interconnected, with the result that they have become more vulnerable to solar effects that disturb current systems in the ionosphere. Disturbed ionospheric conditions are accompanied by the heating of Earth's upper atmosphere, which increases atmospheric density at the altitudes at which spacecraft in low Earth orbit, including the space shuttle and the International Space Station, fly. The increased density worsens drag on spacecraft, causing orbital perturbations that create tracking and operational difficulties and possibly even shortening mission lifetime through accelerated orbital decay.

The bursts of radio waves that accompany large outbursts of solar activity produce unwanted noise and interference in military communications and radar systems. With the burgeoning of the wireless telecommunications age over the last decade, more attention is now being given to the potentially adverse effects of solar radio bursts on these civilian systems as well. The vulnerability of wireless systems to solar activity depends sensitively on numerous design details and is a subject of considerable study.

The lesson learned over more than four decades is that the space environment presents many challenges for engineering and for science if many present-day and future technologies are to be successfully implemented. Moreover, these technical challenges are almost always intertwined with complex policy issues that can take a variety of forms. The following sections examine these issues, both technical and policy-related, in five key areas.

THE NATIONAL SPACE WEATHER PROGRAM

A key function of the National Space Weather Program is to develop processes and policies for monitoring the space weather environment. Technological changes are occurring at a very rapid pace in both large-scale systems and small-scale subsystems that can be affected by solar-terrestrial processes. This is true for both ground- and space-based systems. As an example, the electric power system in the United States is becoming ever more interconnected, through power sharing pools and other mutual arrangements. Such interconnectedness implies that a space-weather-related problem in one region could potentially cascade to a distant region not directly affected by the space weather event. As another example but at the opposite extreme of the size scale, rapid advances continue to occur in microelectronics and the miniaturization of circuits. There is also a continued drive toward the use of commercial off-the-shelf (COTS) components in military programs, including spaceflight programs, and toward including more capabilities in civilian space programs, most importantly in communications satellites. The convergence of these trends (increased miniaturization and use of COTS components) could increase the vulnerabilities of both civilian and military systems.

An important step toward mitigating the adverse effects of solar activity and the space environment on important technological systems was taken by the United States in the mid-1990s when, following advocacy by researchers from academia, government, and industry, the National Space Weather Program (NSWP) was created.[3,4] The focus of this program is on the entire Sun-Earth system, and its goal is "to achieve, within a ten year period, an active, synergistic, interagency system to provide timely, accurate, and reliable space environment observations, specifications, and forecasts." The participating government agencies include the NSF (the lead agency), NASA, and the Departments of Commerce, Defense, Energy, and Transportation. The National Space Weather Program Council, consisting of high-ranking members of the agencies,[5] oversees the NSWP. Following the creation of the NSWP, other studies and initiatives were carried out by several of the agencies, including the *Space Weather Architecture Study* by the DOD.[6]

Some of the agencies are also expanding their contacts and interactions with the external, predominantly university, research communities to tap the expertise residing there and to encourage research activities related to space weather. The DOD Office of Naval Research (ONR) and the Air Force Office of Scientific Research (AFOSR) initiated and now support three multidisciplinary university research initiatives, each of which is a 5-year program focused on developing physics-based assimilation models for the ionosphere, the magnetosphere, and the solar environment. The DOD-sponsored University Partnership for Operational Support, the goal of which is to develop space weather products for near-term use, is centered at the University of Alaska and the Johns Hopkins University.

The National Science Foundation has implemented a dedicated line of resources (augmented with funding from the AFOSR and the ONR) that is allocated by peer review for space weather studies. The agency also supports specialized workshops and activities related to space weather.

NASA has initiated the Living With a Star program to "develop scientific knowledge and understanding of those aspects of the connected Sun-Earth system that directly affect life and society."[7] The program includes four elements: (1) a space weather research network of solar-terrestrial spacecraft; (2) a theory, modeling, and data analysis program; (3) space environment testbeds for flight testing of radiation-hardened systems in the near-

Earth environment; and (4) the development of partnerships with national and international agencies involved with space weather concerns.

NOAA's operational activities for space weather are centered at its Space Environment Center in Boulder, Colorado, where data are used and models are developed to aid in the forecasting and monitoring of solar-terrestrial conditions (see Figure 5.2). The Space Environment Center is the element of NOAA that is involved in defining continuing and new measurements and instruments for its civilian operational mission.

Over the last decade, a number of private companies have also become interested in issues related to space weather. This interest has often arisen as business extensions of related services that these companies had already been supplying to customers. As a result, some of these companies are beginning to provide space weather products to both private and public (including DOD) entities.

Finding: The federal agencies that have important research and/or mission interests in the solar-terrestrial environment are undertaking strong initiatives to establish, nurture, and evolve an effective national program in space weather. There is growing interest in the private sector in the provision of space weather products to both the private and the public sectors. As a result of all of these activities, numerous research and research policy issues have arisen that demand new attention from all parties interested in space weather.

MONITORING THE SOLAR-TERRESTRIAL ENVIRONMENT

Effective monitoring of the space environment requires identification of those research instruments and observations that are needed to provide the basis for modeling interactions of the solar-terrestrial environment with technical systems and for making sound technical design decisions. Considerable progress has been made in the last decade in understanding the effects of the space environment on Earth and its many technological systems. Ongoing and planned research will lead to further advances in the knowledge base. The vast majority of this research is motivated by intellectual curiosity. Not all of it is directly applicable to the products that potential users are calling for, nor should it be. The transfer of the results of basic research to practical use and products is one of the largest conundrums faced by technology managers and researchers. The converse relation, how the numerous puzzles and questions that arise from real-world problems can be used to influence research directions, also remains an issue.

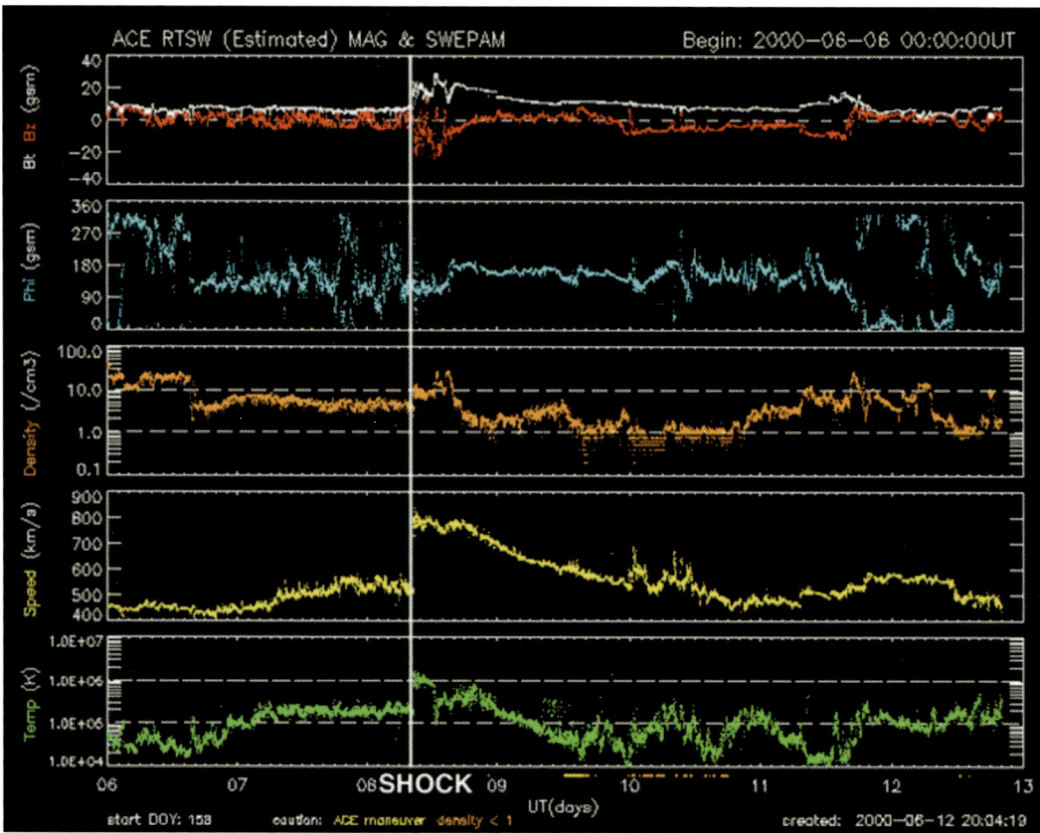

FIGURE 5.2 An example of real-time interplanetary data from the magnetic field experiment (top two panels) and the solar wind plasma instrument (bottom three panels) on NASA's Advanced Composition Explorer (ACE) spacecraft. The data presented here cover a period of 7 days, from June 6 through June 12, 2000, and show a CME-driven interplanetary shock that passed the spacecraft on the morning of June 8 and triggered a moderate geomagnetic storm later that day. The passage of the shock is revealed by the rise in the speed (yellow line), density (orange line), and temperature (green line) of the solar wind and in the magnitude of the interplanetary magnetic field (white line). The strong southward (negative) component (red line) of the interplanetary magnetic field at the shock effects the transfer of solar wind energy into the magnetosphere, which drives the geomagnetic storm. Changes in the overall angle of the interplanetary field relative to the Earth-Sun line are indicated by the blue trace. Positioned at the Lagrangian point L1, 1.5 million kilometers upstream from Earth, ACE can provide approximately an hour's advance information about Earthward-directed solar wind disturbances. Real-time ACE solar wind data are made available to the user community through the Web site of NOAA's Space Environment Center. An upstream solar wind monitor is a critically important asset for both research and space weather applications. Courtesy of the ACE project and the NOAA Space Environment Center.

A unique and nationally important aspect of solar and space physics research continues to be the rapid transition of new understanding of the Sun and the geospace environment to practical applications. This transition is especially important for federal agencies such as the DOD and NOAA. New research instrumentation deployed on the ground or in space often leads to major advances in new data capabilities that can be incorporated into applications and operational systems. Moreover, the continued acquisition of certain kinds of data by some existing facilities is important for operational purposes.

The responsibility for acquiring data on the solar-terrestrial environment is distributed among different agencies. For example, responsibility for maintaining the ground-based magnetometers, whose data are used for operational purposes by NOAA and the private sector, rests with the Department of the Interior. The space-based science platforms that serve the nation's civilian space program are almost exclusively the responsibility of NASA. In both cases, as well as in the case of other facilities that perform virtually routine monitoring functions, the chief beneficiaries are NOAA's operational programs. Transitioning such data acquisition programs and/or their acquisition platforms into operational use requires strong and effective coordination among agencies.

An example of a data acquisition activity that is of critical importance for both scientific and operational purposes and that raises questions of continued availability and of the transition from science to operations is the upstream monitoring of the solar wind and the interplanetary magnetic field. Currently, NASA's Advanced Composition Explorer (ACE) spacecraft, located at the Lagrangian point L1, is providing key solar wind data used in NOAA's operational programs and by private companies for their customers (see Figure 5.2). ACE is currently operating beyond its design life, however, and how such data are to be provided in the future will have to be seriously considered. International intentions and plans for L1 monitors will have to be taken into account as well. The future availability of interplanetary data, whether they come from L1 (these data provide some measure of early warning of disturbed conditions) or near Earth (these data can assess more accurately conditions that might affect Earth), always involves some uncertainty since the scientific peer review process does not normally give routine monitoring a high priority. Continuation of more routine monitoring of the geophysical environment by some ground facilities such as magnetometers and cosmic ray neutron monitors can be problematical as well, even when their usefulness for space weather applications is widely recognized.

Recommendation: NOAA and DOD, in consultation with the research community, should lead in an effort by all involved agencies to jointly assess instrument facilities that contribute key data to public and private space weather models and to operational programs. They should then determine a strategy to maintain the needed facilities and/or work to establish new facilities. The results of this effort should be available for public dissemination.

Recommendation: NOAA should assume responsibility for the continuance of space-based measurements such as solar wind data from the L1 location as well as near Earth and for distribution of the data for operational use.[8]

Dramatic advances in understanding the causative links between CMEs and high-speed interplanetary streams and large geomagnetic disturbances have come from the solar remote-sensing instrumentation on the Yohkoh and SOHO spacecraft, both of which involve international collaborations. These advances have motivated the development of both the Solar Dynamics Observatory (SDO) and the STEREO mission. Imaging of key regions of geospace, such as the auroral zone, is now becoming more routine and therefore more useful for operational applications—for example, for monitoring and following electrojet activity, which can affect ground-based systems. There is now little doubt that imaging of the Sun and of geospace will one day play a central role in operational space weather forecasting.

Recommendation: NASA and NOAA should initiate the necessary planning to transition solar and geospace imaging instrumentation into operational programs for the public and private sectors.

Such transitions of other instrumentation have occurred over the years, from the addition of charged-particle sensors to operational weather satellites some years ago to the more recent incorporation of new concepts such as solar x-ray spectral and imaging data.

THE TRANSITION FROM RESEARCH TO OPERATIONS

An important task facing the space weather community during the coming decade will be to establish, maintain, and evolve mechanisms for the efficient transfer of new models of the solar-terrestrial environment into the user community. The mere acquisition of new knowledge, whether in the form of data sets or theoretical insights, is insufficient for practical uses.

Means must be established for transitioning the new knowledge into those arenas where it is needed for design and operations. Creative, cutting-edge research in modeling the solar-terrestrial environment has been under way in the research community for many years. Insofar as possible, these efforts use contemporary theoretical knowledge, current data sets, and, for some models and when data are available, real-time data for model validation and operational use. In recent years, proprietary models have also been introduced by private companies for use in both the public and private sectors.

Under the auspices of the National Space Weather Program, models thought to be potentially useful for space weather applications can be submitted to the Community Coordinated Modeling Center (currently located at the NASA Goddard Space Flight Center) for testing and validation. Following validation, the models can be turned over to either the U.S. Air Force or the NOAA Rapid Prototyping Centers, where they are used for the objectives of the individual agencies. DOD takes the research models and produces tailored products for its specific needs, while NOAA forecasts and specifies the solar-terrestrial environment.

In many instances, the validation of research products and models differs in the private and public sectors. In the private sector, validation generally occurs via the marketplace, when the customer pays for and uses the model or other product. The continued use and payment by the customer (government or private) tells the vendor that value has been added to the customer's business. In contrast, publicly funded research models and system-impact products are usually placed in an operational setting with limited validation.

Recommendation: The relevant federal agencies should establish an overall verification and validation program for all publicly funded models and system-impact products before they become operational.

Over the years, the scientific community has developed many different models of the near-Earth space environment that could have practical applications. More recently, space environment models have been developed by commercial interests as well, with details of these private models often being proprietary. Public, nonclassified models now cover all of the near-Earth space domains, including the solar wind, magnetosphere, radiation belts, ionosphere, and thermosphere. Typically, for each domain, there are several different models developed by different individuals.

Recommendation: The operational federal agencies, NOAA and DOD, should establish procedures to identify and prioritize operational needs, and these needs should determine which model types are selected for transitioning by the Community Coordinated Modeling Center and the Rapid Prototyping Centers. After the needs have been prioritized, procedures should be established to determine which of the competing models, public or private, is best suited for a particular operational requirement.

DATA ACQUISITION AND AVAILABILITY

Developing successful space weather mitigation strategies involves the ability to predict space weather effects on specific technological systems as well as to predict space weather in general; it also requires a knowledge of extreme space environmental conditions. An oft-stated goal of researchers and those who use the results of the research is the ability to reliably predict the effects on geospace and specific technological systems following an event on the Sun. The "transfer functions" that relate a given solar observation to the effects on a specific technological system are largely unknown. For example, the vulnerability of an interconnected electrical power grid may depend on the location of a specific portion of the system relative to the instantaneous location of a suddenly changing electrojet current as well as on the conductivity of the ground beneath the system element.

The scintillations that might be encountered by a given radio transmission path can be highly variable depending on the state of the ionospheric plasmas and current systems in the path's line of sight. Ionospheric current systems can be followed now in near-real time with appropriate instrumentation. However, forecasts of their movements and intensities and plasma properties from the data available are still primitive, to say nothing of the forecasts that might be derived from observations of conditions at the Sun and from subsequent solar wind measurements at L1. Related to this challenge is the need to identify existing databases that might provide a perspective on extreme conditions in the Sun-Earth system. Designing for operations under extreme conditions could be one way of relaxing space weather predictive requirements for implemented operational systems.

During the coming decade, gigabytes of data per day could be available for incorporation into physics-based data assimilation models of the solar-terrestrial environment and into system-impact codes for space weather forecasting and mitigation. The situation in solar and space physics regarding data assimilation will then begin to resemble the state of data assimilation that exists today in meteorology. These solar and space physics data will be available for public codes, for national security codes, and for proprietary commercial codes. At present, there are several obstacles that hinder data assimilation. First, DOD traditionally uses its own data (which are often not available to outside users) and has only recently begun to use data from other agencies and institutions. Therefore, many data sets are not available for use by the publicly funded or commercial vendors that design products for DOD or would like to compete for product opportunities. In addition, the usefulness of data assimilation techniques will be unduly limited, since only a small fraction of the data is likely to be available in real time.

> **Recommendation: DOD and NOAA should be the lead agencies in acquiring all the data sets needed for accurate specification and forecast modeling, including data from the international community. Because it is extremely important to have real-time data, both space- and ground-based, for predictive purposes, NOAA and DOD should invest in new ways to acquire real-time data from all of the ground- and space-based sources available to them. All data acquired should contain error estimates, which are required by data assimilation models.**

Implementation of this recommendation will require good coordination among agencies (see Chapter 7).

When designing ground- and space-based systems, engineers are typically interested in space climate and not space weather. Needed are long-term averages, the uncertainties in these averages, and values for extreme conditions. The engineering goal is to design a system to be immune to space weather effects as much as is feasible. That is, the space environment should be removed from the equation; any further space weather issues that might arise can be dealt with separately. When climatological models are developed, extreme conditions are either ignored or not properly represented because there are too few data points to justify including them in a statistical database.

> **Recommendation: A new, centralized database of extreme space weather conditions should be created that covers as many of the relevant space weather parameters as possible.**

A possible location for the database is within NOAA, at its Space Environment Center or its National Geophysical Data Center. The database should primarily contain measurements, and resources will be needed to search existing archives for the extremes for all relevant environmental parameters. The outputs of physics-based model runs for extreme conditions, appropriately documented and annotated, should be a part of the database.

THE PUBLIC AND PRIVATE SECTORS IN SPACE WEATHER APPLICATIONS

Both the government and private industry are involved in acquiring, assessing, and disseminating information and models related to the solar-terrestrial environment in the context of its relevance for technological systems. Therefore, it is important to determine the appropriate roles for each sector in space-weather-related activities. To date, the largest efforts to understand the solar-terrestrial environment and apply that understanding for practical purposes have been mostly publicly funded and undertaken by government research organizations, universities, and some companies. Recently, some private companies both large and small have been devoting their own resources to the development and sale of specialized products that address the design and operation of certain technical systems that can be affected by the solar-terrestrial environment. Such private efforts often use publicly supported assets (such as spacecraft data) as well as proprietary instrumentation and models. A number of the private efforts use proprietary system knowledge to guide their choices of research directions.

While some publicly and privately funded efforts are beginning to compete with one another to perform similar tasks, all parties recognize that synergistic benefits can occur through continuing collaboration and the clear definition of responsibilities among complementary organizations. Still, private-sector policies on such matters as data rights, intellectual property rights and responsibilities, and benchmarking criteria can be quite different from the policies that apply for publicly supported space-weather-related activities, including those performed at universities. Thus, transitioning knowledge and models from one sector to another can be fraught with complications and requires continued attention and discussion by all interested entities. Similar issues arise with regard to provision of public and private value-added meteorological data and data products,[9,10] and lessons learned from the meteorology community can potentially be utilized as the national space weather effort grows, evolves, and matures.

Recommendation: Clear policies should be developed that describe government and industry roles, rights, and responsibilities in space

weather activities. Such policies are necessary to optimize the benefits of the national investments, public and private, that are being made.

The implementation of this recommendation should be led by NOAA and DOD, together with one or two commercial entities. The public sector might focus, for instance, on acquiring the real-time data needed for reliable forecasts as well as on the production of generic forecast information, while the private sector might focus on value-added, system-impact products, including forecast products that would be tailored to specific systems or missions. These policies could be developed in a variety of ways, one being to convene a study group or a workshop under the auspices of a third party.

NOTES

1. Lanzerotti, L.J., et al., Engineering issues in space weather, in *Modern Radio Science*, M.A. Stuchly, ed., John Wiley & Sons, Inc., Hoboken, N.J., 1999.
2. Lanzerotti, L.J., Space weather effects on technologies, in *Space Weather*, Geophysical Monograph 125, P. Song, H.J. Singer, and G.L. Siscoe, eds., American Geophysical Union, Washington, D.C., 2001, pp. 11-22.
3. Office of the Federal Coordinator for Meteorological Services and Supporting Research, *The National Space Weather Program: The Strategic Plan,* FCM-P30-1995, Silver Spring, Md., August 1995.
4. Office of the Federal Coordinator for Meteorological Services and Supporting Research, *The National Space Weather Program: The Implementation Plan*, 2nd Edition, FCM-P31-2000, Silver Spring, Md., July 2000.
5. Robinson, R.M., and R.A. Behnke, The U.S. National Space Weather Program: A retrospective, in *Space Weather*, Geophysical Monograph 125, P. Song, H.J. Singer, and G.L. Siscoe, eds., American Geophysical Union, Washington, D.C., 2001, pp. 1-10.
6. National Security Space Architect, *Space Weather Architecture Study Transition Strategy*, March 1999. Available online at <http://schnarff.com/SpaceWeather/PDF/Reports/P-IIB/02.pdf>.
7. Withbroe, G.L., Living With a Star, in *Space Weather*, Geophysical Monograph 125, P. Song, H.J. Singer, and G.L. Siscoe, eds., American Geophysical Union, Washington, D.C., 2001, pp. 45-51.
8. For example, a NOAA-Air Force program is producing operational solar x-ray data. The Geostationary Operational Environmental Satellite (GOES) Solar X-ray Imager (SXI), first deployed on GOES-M, took its first image on September 7, 2001. The SXI instrument is designed to obtain a continuous sequence of coronal x-ray images at a 1-minute cadence. These images are being used by NOAA's Space Environment Center and the broader community to monitor solar activity for its effects on Earth's upper atmosphere and the near-space environment.
9. National Research Council, *A Vision of the National Weather Service: Roadmap for the Future*, National Academy Press, Washington, D.C., 1999.
10. Serafin, R.L., et al., Transition of weather research to operations: Opportunities and challenges, *Bulletin of the American Meteorological Society* 83, 377-392, 2002.

6

Education and Public Outreach

The committee's consideration of issues related to education and outreach was driven by two main concerns:

• Can the education system provide a sufficient number of scientists trained in solar and space physics to carry out the national research program outlined in this report for the next decade?
• How can solar and space physics contribute to the national effort to enhance education in science and technology?

Dealing with issues in these two areas can, the committee believes, provide the most leverage for the future.

To address the specific need for people trained in solar and space physics, the committee concentrated on colleges and universities, where declining enrollments in undergraduate degree programs in physics and Earth sciences are leading to a shortfall at the base of the pipeline for future researchers, instrument developers, faculty, and mentors in solar and space physics. It considered ways to enhance the quality of education at the dozen or so colleges and universities where solar and space physics currently has a strong presence and looked at approaches to attracting a diverse student population to the field and to encouraging the expansion of solar and space physics to a larger fraction of the nation's institutions of higher education. The committee focused mostly on issues related specifically to undergraduate education, believing that it is here that research and teaching in solar and space physics can have the greatest impact over the next decade.

To address the broader issue of contributions by solar and space physics to science literacy and appreciation, the committee focused on undergraduate physics and astronomy general education courses, which influence a significant number of potential school teachers at all levels, as well as on programs that support the involvement of solar and space physics researchers with institutions such as local schools and museums.

The committee believes that solar and space physics research should continue to make substantial contributions to K-12 education, to informal education of the public, and to public outreach. Solar and space physics can deepen people's awareness of the excitement of space exploration, the beauty of auroral displays, and the drama of massive eruptions on the Sun. And as described in Chapter 5, the practical applications of solar and space physics readily illustrate the societal relevance of understanding the solar system. Solar and space physics can be a valuable tool for motivating and educating students, for informing the public, and for illustrating important lessons of physical science (Figure 6.1).

The strengthening of science and technology education is an important national goal whose achievement would bolster the country's workforce in science and engineering and also ensure a citizenry that is able to cope with and understand the technical forces that are shaping the contemporary world.[1]

EDUCATING FUTURE SOLAR AND SPACE PHYSICISTS

Augmenting the Faculty

When the newest results from spacecraft exploration of the Sun and of Earth's environment in space are reported, even the youngest students express great interest and enthusiasm. Yet in part because of its relatively short history, solar and space physics appears only adventitiously in formal instructional programs. Because it is mainly in the colleges and universities that new science is presented to future generations, new issues are debated, and interdisciplinary approaches are developed,[2] it is vital that solar and space physics develop and maintain a strong presence in our colleges and universities.

Solar and space physics is highly interdisciplinary, with faculty and graduate education often split among departments of physics, mathematics, geophysics, astronomy, electrical and aerospace engineering, and Earth and atmospheric sciences. Currently, only a handful of institutions offer specific undergraduate courses, much less concentrations, in solar or space physics. Moreover, unlike introductory astronomy or geology, which are prominent in K-12 and college science instruction, solar and space physics is rarely a part of current curricula. Thus, many who graduate with degrees in the

FIGURE 6.1 Public outreach by the solar and space physics community spans a wide range of activities through different media and at a variety of venues. Examples are (a) the IMAX movie *SolarMax*, playing at theaters across the nation, <http://www.solarmovie.com/>; (b) a plasmasphere, part of the Electric Space exhibit that has toured science museums, <http://www.spacescience.org/Outreach/TravSciExhibits/ESpaceExhibitProject/>; (c) information about space weather distributed via the Web and compact disks, <http://earth.rice.edu/connected/space_weather.html>; and (d) a press release of auroral movies from the Polar spacecraft, <http://www.gsfc.nasa.gov/topstory/20011025aurora.html>.

---

physical sciences have little knowledge of the space environment that envelops our planet.

At the same time, solar and space physics has continued to become an ever greater element of our national research portfolio. Indeed, as summarized in this survey, recent advances in understanding in solar and space physics have been enormous. The number of active researchers has continued to grow along with research opportunities, and the ambitious, and quite achievable, research program laid out in this survey report would further enhance understanding of the Sun in many key areas, as well as of the space environments of Earth and the other planets, over the next decade. But success in this research requires a strong national cadre of young and expert solar and space physics scientists who can participate in research (particularly instrument development) whether at universities, in industry, or at the national laboratories. Thus the currently dwindling pipeline for solar and space physics researchers is a concern. For the past 10 years the number of bachelor's degrees awarded in physics has declined by 20 percent (at the same time as the total number of bachelor's degrees has increased by 20 percent).[3] The number of doctorates granted in physics has held steady only because of the increased participation of foreign students; the number of physics doctorates awarded to U.S. students has continued to drop. Strong measures of diverse types are required to attract students to the relevant programs and maintain effective research at universities.[4]

Various approaches could be used to encourage colleges and universities, particularly the top-tier research universities, to include solar and space physics topics as an integral part of their physical science curricula and to foster continuing strong research in the field. A healthy presence for solar and space physics in academia would require additional faculty members to guide student research (both undergraduate and graduate), to teach solar and space physics graduate programs, and to integrate topics in solar and space physics into basic physics and astronomy classes. A program that provides matching funds would give academic institutions a critical incentive to recruit solar and space physics faculty. Adding solar and space physics faculty to institutions serving minorities could allow these institutions to expand their research agendas and serve as recruiting grounds for solar and space physics. Augmentation of university faculty in solar and space physics is essential for the support of a strong national solar and space physics research program in the coming decade.

Recommendation: The NSF and NASA should jointly establish a program of "bridged positions" that provides (through a competitive process) partial salary, start-up funding, and research support for four new faculty members every year for 5 years.

Accurate statistics on solar and space physics demographics do not exist. If one makes a rough estimate that there are 100 solar and space physics faculty in tenure-track positions at universities, the recommended support for a total of 20 new faculty lines over the next decade represents an increase of ~20 percent. Each appointment would be made in accordance with an academic institution's normal appointment and tenure-track policies, with the federal agencies providing a portion of the salary and research funds for 3 to 5 years (depending on the seniority of the recipient) to help initiate courses in the field and to start a research group. After 5 years, the position would be reviewed under the institution's regular assessment system. If the assessment proved satisfactory, and if—as expected—the research group had become self-sustaining, the academic institution would then assume full funding for the faculty member. Examples of NASA-supported faculty positions in the past 15 years include those at the University of New Hampshire, Montana State University, and Utah State University. Joint NSF and NASA support for this program is recommended because solar and space physics is of significance to both agencies.

Supporting Summer Institutes and Distance Education

The decadal research program presented in this survey emphasizes the coupled complexity of the interrelated system of the Sun and the heliosphere and the interaction of the solar wind with the planets. Yet the full scope of the relevant scientific background is rarely covered at the graduate level. As pointed out above, the field of solar and space physics spans diverse subject areas that are rarely found in a single academic department and indeed, solar and space physics is taught at only a dozen or so U.S. universities in courses that often cover only a limited part of the field.

Summer institutes have proven successful in exposing graduate students and postdoctoral researchers to a wider range of topics than is typically available at their home institutions. Such institutes assemble graduate students, postdoctoral researchers, and experienced space physicists for 2 weeks to 3 months of concentrated study. Attendees receive tutorials in the underlying physics, learn about the latest research across the entire breadth of solar and space physics and/or in specialized topics, and become familiar with the connections across the solar and space physics domain. Several entities, including the Santa Barbara Theory Institute, the NSF Geospace Environment Modeling (GEM) program, and the North Atlantic Treaty Organization, have supported such solar and space physics training events. Summer institutes in space physics also continue to be a very successful component of annual week-long workshops at the NSF-sponsored GEM meeting and the Coupling, Energetics, and Dynamics of Atmospheric Regions (CEDAR) meeting; at the University of Alaska Ionospheric Modification summer school; and at the Center for Integrated Space Weather Modeling at Boston University, which organizes a 2-week summer program for 24 students every year.

Education in solar and space physics during the academic year could be considerably enhanced if the latest advances in information technology were exploited to provide distance learning for both graduate students and postdoctoral researchers. Distance learning courses involving several faculty from a number of institutions could be offered on the Web, either as informal, asynchronous learning opportunities or as formal, synchronous courses.[5] Such offerings would substantially expand the reach and the educational value of the expertise that currently resides at a limited number of institutions. Linking distance education to space physics summer courses would multiply its value, since experience demonstrates that remote learning can be greatly enhanced when it is coupled with face-to-face sessions.[6] Such a program would bring a broader range of solar and space physics educational opportunities to colleges and universities across the nation.

Finding: Summer institutes play a vital role in providing both depth and breadth in the space physics education of graduate students and junior researchers.

Recommendation: The NSF and NASA should jointly support an initiative that provides increased opportunities for distance education in solar and space physics.

Providing Undergraduate Research Opportunities in Solar and Space Physics

Solar and space physics projects for undergraduates offer students a chance to be involved in research, from designing and building hardware to gathering and analyzing data. Ranging from student-based missions—e.g., the Student Nitric Oxide Explorer, designed, built, and operated at the University of Colorado with the participation of more than 100 students—to single-student projects in data analysis, theory, or experimentation, such research opportunities can help students directly realize the excitement of scientific exploration (Figure 6.2). At Augsburg College, for example, a student analyzed data from an array of NSF-supported instruments in the Antarctic; at Middlebury College, a student worked through the algebra of an analytical model of solar prominences; at the University of California at Los Angeles, students helped to archive Galileo magnetometer data; and at Williams College, students joined an expedition to view a solar eclipse and study the solar corona. Summer or part-time internships in industry or government research labs also enable students to experience the research

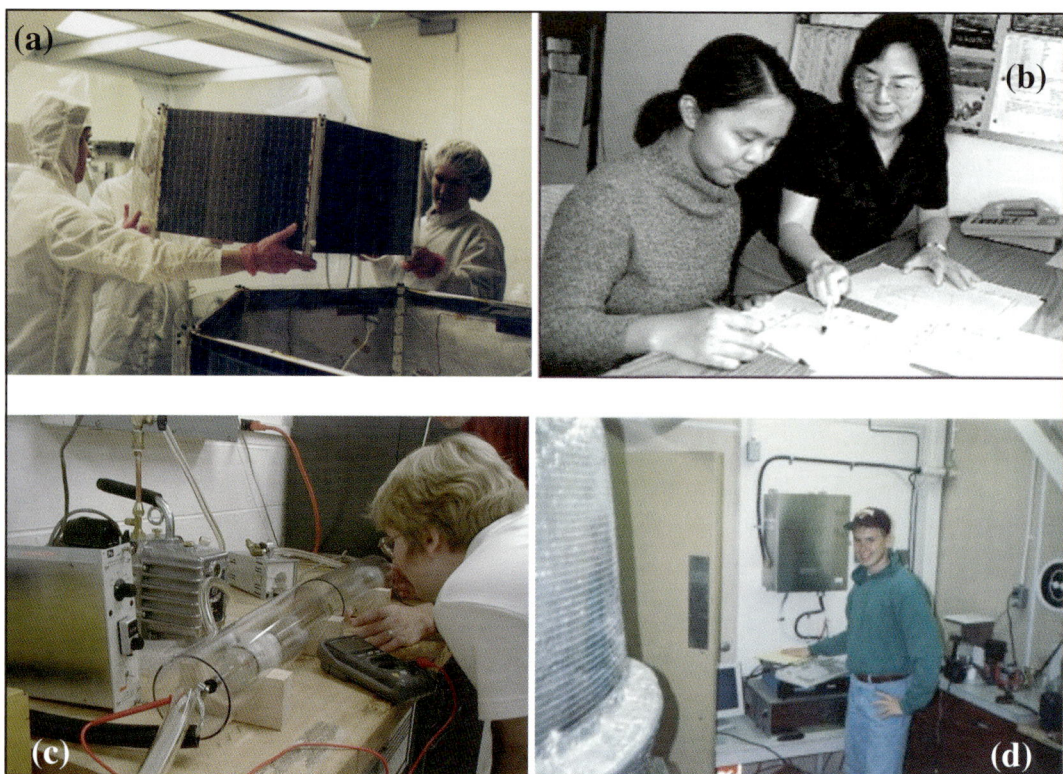

FIGURE 6.2 Involvement of undergraduate students in solar and space physics research has proved to be an important factor in recruiting and retaining students as physics and Earth science majors. (a) Students at the University of Colorado built and operated the Student Nitric Oxide Explorer, <http://lasp.colorado.edu/snoe/>; (b) an individual student working on a theoretical problem with a professional scientist during a Significant Opportunities in Atmospheric Research and Science summer program, <http://www.ucar.edu/soars/dirindex.html>; (c) a student participating in an experiment at the Princeton Plasma Physics Laboratory summer institute for high school teachers, <http://science-education.pppl.gov/SummerInst/index.htmL>; and (d) an Augsburg College undergraduate installing a magnetometer in northern Canada, <http://www.augsburg.edu/physics/>.

environment and to become actively involved in space missions or ground-based projects. At Goddard Space Flight Center, for example, a summer intern analyzed Mars Global Surveyor magnetometer data; at NASA's Jet Propulsion Laboratory, a California Institute of Technology student worked on Ulysses data between classes; and at Lockheed Martin and Lucent Technologies students are analyzing solar data from TRACE or ground-based telescopes.

Undergraduate research programs have contributed effectively to enhancing the recruitment and retention of science majors, as seen, for example, in the successful efforts of some physics departments to increase the number of graduating majors (see sidebar, "Degrees Awarded in Undergraduate Physics Programs—Gains and Losses"). Initiatives aimed at attracting a broader cross section of students to solar and space physics are also vital (see sidebar, "Diversity in Space Physics"). On-campus undergraduate research and off-campus research internships are valued greatly because of the hands-on experience they provide. They also offer students the opportunity to work in a team or group environment, make decisions that contribute to the success or failure of a project, and take responsibility for and feel the satisfaction of being creative. Employers and graduate schools look for this experience in applicants.[7,8]

The NSF's successful Research Experiences for Undergraduates program supports ~15 supplemental grants per year in the solar and space physics area. A simple letter to the program officer specifies how the money (usually a few thousand dollars per student) will be spent, and the decision to fund the request is made by the program officer. At present there is no comparable NASA program for undergraduate research.

Finding: NSF support for the Research Experiences for Undergraduates program is valuable for encouraging undergraduates in the solar and space physics research area.

A substantial expansion of research opportunities for undergraduates in solar and space physics is a means not just of enhancing the training of future scientists in this field but also of contributing to a technically trained workforce.

Recommendation: NASA should institute a specific program for the support of undergraduate research in solar and space physics at colleges and universities. The program should have the flexibility to support such research with either a supplement to existing grants or with a stand-alone grant.

DEGREES AWARDED IN UNDERGRADUATE PHYSICS PROGRAMS—GAINS AND LOSSES

Statistics for 2000 compiled by the American Institute of Physics on the 770 physics departments at U.S. universities show that the number of degrees awarded in physics has continued to decline sharply, dropping by 20 percent over the past 10 years. During the same period, the total number of bachelor's degrees granted in all fields increased by 20 percent.[1] The fraction of women among recipients of physics degrees remains at less than 20 percent and is increasing at a slower rate than the fraction of women earning degrees in engineering and chemistry.[2] Some physics departments are struggling with understanding the causes of these declines and are searching for solutions, as shown by the following excerpts from the report of the Conference on Building Undergraduate Physics Programs for the 21st Century, held in Arlington, Va., October 2-4, 1998. Some have succeeded in implementing reforms that have had positive effects on enrollment.

> Ten college and university departments presented "case studies" of their responses to the challenges faced by their physics departments.... The departments had taken different paths, but many programs had overlapping themes. Some promoted mentoring and recruiting of undergraduates and involving undergraduates in research as soon as possible. Others had made major modifications in their curricula to promote double majors or to allow students to take courses in engineering specialties to make themselves more marketable. Still others had made major changes in the way they teach their introductory courses to make them more attractive and useful to students. Bob Ehrlich, Professor of Physics and former chair at George Mason University, analyzed data collected from physics departments that had seen a significant change in the numbers of majors completing their programs. He analyzed the situations for those who experienced an increase (7 "Big Gainers") and those who had seen a significant decline (28 "Big Losers") in the numbers of majors they graduate. He found it significant that the "Big Losers" blame their declines mainly on external factors such as increased competition from other departments, changes in student preparation or skewed statistics. Few of the department chairs see themselves or their fellow physics faculty members as the problem, although one chair mentioned aging faculty as a significant factor in the decline.
>
> In contrast, the seven "Big Gainers" had implemented reformed curricula, particularly in the introductory courses. They had focused on increased recruitment efforts and adopted mentoring programs to increase retention. These departments encouraged early involvement of undergraduates in research, and provided extensive advising, and community-building within the department. Some of them cited grants as being critical to their success. Nearly all of the Big Gainers had introduced flexible, multiple-track majors that allowed their students time to take courses outside the physics department—in engineering or computer science, for example.

---

1. Nickholson, S., and P. Mulvey, *American Institute of Physics Report R-394-7*, September 2001.

2. Ivie, R., and K. Stowe, *American Institute of Physics Report R-430*, June 2000.

DIVERSITY IN SPACE PHYSICS

The solar and space physics community should seek ways to avail itself of the talents and skills of a large and increasing segment of society, individuals from underrepresented groups. The need to attract a diverse population to the field is articulated in the Diversity Strategy of the American Geophysical Union:

> The Earth and space sciences are in danger of losing a significant portion of the workforce necessary to ensure its future. Evidence for this problem includes:
>
> • The aging population of scientific professionals nearing retirement comprises the largest proportion of the present Earth and space scientists.
> • A thirteen percent decline in graduate enrollment occurred within the Earth and space sciences during the 1990's.[1]
> • The numbers of white males, the largest demographic community within the Earth and space sciences, receiving bachelors degrees in the geosciences have decreased by nearly 80% over the past quarter century.[2] Thus, the traditional base of future Earth and space scientists in the US is shrinking. And,
> • Over the last two decades, the numbers of Earth and space science academic programs, particularly at post-secondary levels, and total academic science majors in the United States have declined.
>
> The 21st century demographics of the US population in grades K-12, i.e., the future scientists of America, are shifting rapidly. Minority populations have had the greatest proportional increase within the United States during the decade of the 1990's. Presently, racial and ethnic minorities, women, and persons with disabilities are not replacing the potential workforce shortfall. This is despite the fact that the percentage of ethnic and racial minorities in the resident US population is ~40% of the future talent pool, i.e. elementary school students.[3] Students from minority groups are not choosing geoscience careers for a variety of reasons, not all of which are fully understood.
>
> . . . It is essential that new strategies for educating, recruiting, and retaining geoscientists from currently under-represented populations be developed in order to fill this future workforce shortfall. The potential ramifications of this situation—for individual investigators seeking students to fill classes or work in their research programs, for institutions looking to replace faculty and researchers, for the larger community looking to the public for continued research funding . . . could be crippling. Therefore, the challenge is to identify, promote, and implement effective strategies that increase diversity within the Earth and space sciences.

The federal agencies supporting solar and space physics recognize these issues and have set up initiatives to address them. The NSF's Diversity Initiative program has offered grants to a variety of universities and organizations such as the Society for the Advancement of Chicanos and Native Americans in Science. In the summer of 2000, NASA's Office of Space Science launched its Minority University Initiative, which has made available to institutions serving minorities funds for a wide range of programs, such as new space science courses or degree programs, public education, and outreach efforts. NOAA has established a diversity initiative aimed at supporting NOAA-related science research targeted at minority-serving universities, and it issued a request for proposals in 2002.

---

1. NSF, Data Brief: *Growth Continued in Graduate Enrollment in Science and Engineering Fields*, NSF 01-312, 2001; NSF, *Graduate Students and Postdoctorates in Science and Engineering: Fall 1999*, NSF 01-315, 2001.
2. American Geological Institute, *Guide to Geoscience Departments*, 2001.
3. United States Census Bureau, *NP-D1-A: Projections of the Resident Population by Age, Sex, Race, Hispanic Origin, and Nativity: 1999-2100*, 2000.

ENHANCING EDUCATION IN SCIENCE AND TECHNOLOGY

Curiosity about our surroundings and about the universe beyond our immediate horizon has driven intellectual musings and scientific exploration over the millennia. The space age not only has given birth to the new research discipline of solar and space physics, but also has sparked intense public interest in the space environment and space exploration. Efforts to expand understanding, driven initially by curiosity, have found important practical applications that, as related in Chapter 5, rely on satellite information systems and space-based communications. Both the science of solar and space physics and the societal implications of solar and space physics phenomena should be conveyed to students and the public through a diverse program of educational activities.

Solar and Space Physics in Basic Undergraduate Instruction

The inclusion of solar and space physics topics in instructional programs that reach a student audience beyond physical science and engineering majors is another important educational goal. In the basic physics and introductory astronomy courses taken by students in many fields to fulfill the science requirement of a general education component and taught at just about every university and college across the nation, solar and space physics topics and issues are often treated superficially, if at all. Solar and space physics has much to contribute to the science education of these large audiences—estimated by textbook publishers to number approximately 250,000 students at any one time in introductory astronomy courses, for example—which include science and engineering majors along with nonscience majors.

While basic physics courses are often required for science and engineering majors, there are well-documented problems with the manner in which Physics 101 is taught at many universities and colleges (see sidebar above, "Degrees Awarded in Undergraduate Physics Programs—Gains and Losses"). These include large classes, underqualified instructors, a mechanical approach to learning, and a competitive rather than a collaborative culture. All of these factors have been shown to reduce the motivation of even capable student engineers and physicists, as well as nonscience majors.[9,10] The solar and space physics community could contribute to quality basic undergraduate instruction in a number of ways:

- It could develop learning materials that use examples from space science to illustrate fundamental concepts such as energy, magnetism, and radiation.
- It could provide less experienced instructors with quality materials that are tied to the curriculum, easy to use, self-explanatory, and readily incorporated into collaborative learning activities.
- It could develop Web-based interactive tools that students can use in class or at home in alternative approaches to learning (rather than passively listening to a lecture).
- It could harness the excitement of space exploration through the use of stimulating concepts, events, and phenomena with social relevance—such as auroral images, magnetic weather on the Sun, the latest data from planetary probes, and predictions of radiation fluxes being experienced by astronauts in the International Space Station.

Directed at improving the quality of teaching and learning, these sample approaches follow principles recommended in recent NRC studies[11] and prevalent in the physics education literature.[12]

Solar and space physics can also contribute significantly to the impact of introductory science courses that are taught across the country to strengthen science literacy for nonscientists. Topics such as the Sun as a star, Earth as a planet, and the space environment surrounding Earth and other solar system objects provide opportunities to illustrate basic physical principles with relevant examples (see sidebar, "Space Physics Topics in Introductory Science Courses").

SPACE PHYSICS TOPICS IN INTRODUCTORY SCIENCE COURSES

Introductory Physics

- Magnetic and electric fields
- Charged particle motions, currents
- Plasmas
- Atomic physics—ionization, excitation, radiation, recombination

Introductory Astronomy

- Sun and stars (interior, atmosphere, corona, solar wind)
- Solar variability
- Planetary magnetic fields (implications for interiors, surfaces, and atmospheres)
- Terrestrial space weather, auroras

For many K-12 teachers who take versions of Physics 101 or Astronomy 101 during their preservice education, these courses are their last formal contact with science. Enriching the experience of future teachers with stimulating and relevant material and inquiry-based learning could have a deep, long-term impact on science education in schools. The coupling of undergraduate education to the enhancement of teacher preparation is a major theme of several recent studies.[13]

While some physics and astronomy textbooks are beginning to use images and examples from solar and space physics, modern technology offers a potentially more interactive learning experience (Figure 6.3). Movies of the solar corona from the TRACE spacecraft bring alive concepts of magnetic fields and plasmas. Bright auroral displays dramatically illustrate atomic processes. The IMAGE movies make magnetospheric variability evident. Furthermore, interactive tools allow students to explore concepts such as electric and magnetic fields, the motion of charged particles, and the excitation and radiation of atmospheric gases as they watch the effects of changing parameters in simple graphical models. These interactive, inquiry-based learning activities are proving to be effective in developing long-term understanding of concepts, particularly for students whose levels of achievement in past science courses have tended to be low.[14]

Some of the ingredients of solar and space physics-based instructional materials for introductory physics and astronomy courses already exist. Images, data, and explanations of solar and space physics material appear on many Web sites; faculty at universities around the world also have developed curricular materials for their own courses. Such material should be presented as part of a process of inquiry rather than as an ensemble of facts or knowledge. The solar and space physics community could contribute significantly to a nationwide science literacy program by organizing existing Web resources aimed at audiences with different levels of sophistication, developing additional modules, and providing sample problems in solar and space physics tied to topics in basic physics or astronomy curricula.

Similar tools could be usefully adapted either for more advanced undergraduate courses (e.g., physics courses for majors) or, through collaboration with teachers, for precollege education (discussed also in the next section). For example, instructional materials developed for an introductory astronomy course for nonscientists are often just as appropriate for science-gifted students in middle and high school. Furthermore, with a little repackaging, Web-based material that is developed for formal classes could prove

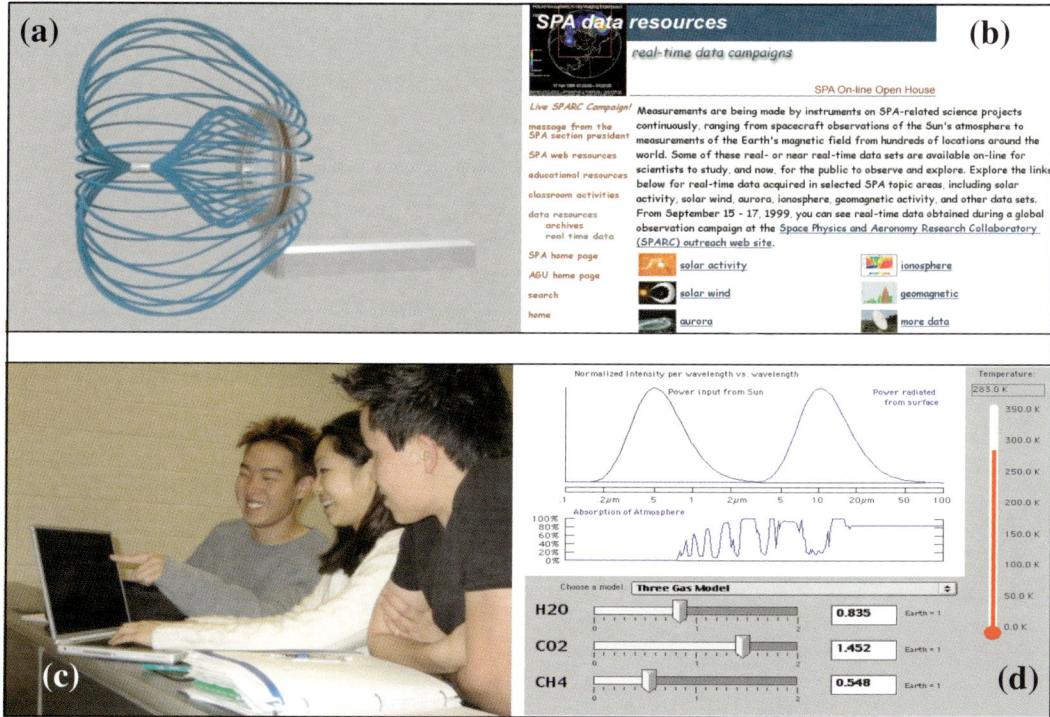

FIGURE 6.3 Solar and space physics teachers are using information technology to develop interactive tools that enhance learning. For example, (a) animations and java applets illustrate experiments in freshman electricity and magnetism at the Massachusetts Institute of Technology, <http://caes.mit.edu/research/teal/>; (b) *Windows to the Universe* provides classroom activities involving links to real-time solar and space physics data, <http://www.windows.ucar.edu/openhouse/data_realtime.html>; (c) University of Colorado students use Web-based tools in an introductory astronomy course, <http://cosmos.colorado.edu/tools>; and (d) a java applet allows students to explore the effect on Earth's temperature of changing the amount of greenhouse gases in the atmosphere, <http://solarsystem.colorado.edu>.

valuable in informal education settings (such as lifelong learning via the Web or in museums and planetariums as well). Solar and space physics researchers will have to team with experienced educators to ensure that such educational products are effective and meet the appropriate curriculum standards (either college or precollege level).

Recommendation: Over the next decade NASA and the NSF should fund groups to develop and disseminate solar and space physics educational resources (especially at the undergraduate level) and to train educators and scientists in the effective use of such resources.

One way to implement this recommendation would be to fund, by peer review, as many as three groups for 3 to 5 years each. A resource development group might be envisioned as a collaborative team comprising solar and space physicists, experienced college teachers, and, when materials are to be adapted for use in middle and high school, curriculum specialists. Such a group could also include experts in the development and assessment of educational materials. The emphasis should be on producing solar and space physics-related materials that can be used broadly (rather than on supporting educational experiments focused at individual institutions) and at multiple levels. Workshops for instructors would encourage effective use of the materials. The committee estimates that the yearly cost of a resource development group would be about $500,000. Such a program could have a substantial impact on science education at the undergraduate level over the next decade.

Solar and Space Physics in K-12 Education and Public Outreach

National Science Education Standards

Solar and space physics also has much to offer in enriching K-12 education (Figure 6.4). The NRC's *National Science Education Standards*, a set of guidelines developed to help teachers and administrators enhance the quality of science education across the nation, covers process and style of education in the sciences and technology as well as specific content.[15] Physical science and Earth and space science are two of the eight subject areas recommended for inclusion in the K-12 science curriculum (see sidebar, "National Science Education Standards").

Particularly in the curriculum for grades 6 through 12, solar and space physics can provide unique illustrations and applications for topics in physical science and Earth and space science and can furnish examples for such content standards areas as science as inquiry, science and technology, and science in personal and social perspectives. Few teachers have formal training in Earth and space science. Individuals in the solar and space physics research community can contribute by becoming involved in their local school district, consulting in the development of appropriate curriculum materials, reviewing the content of such materials for accuracy, and,

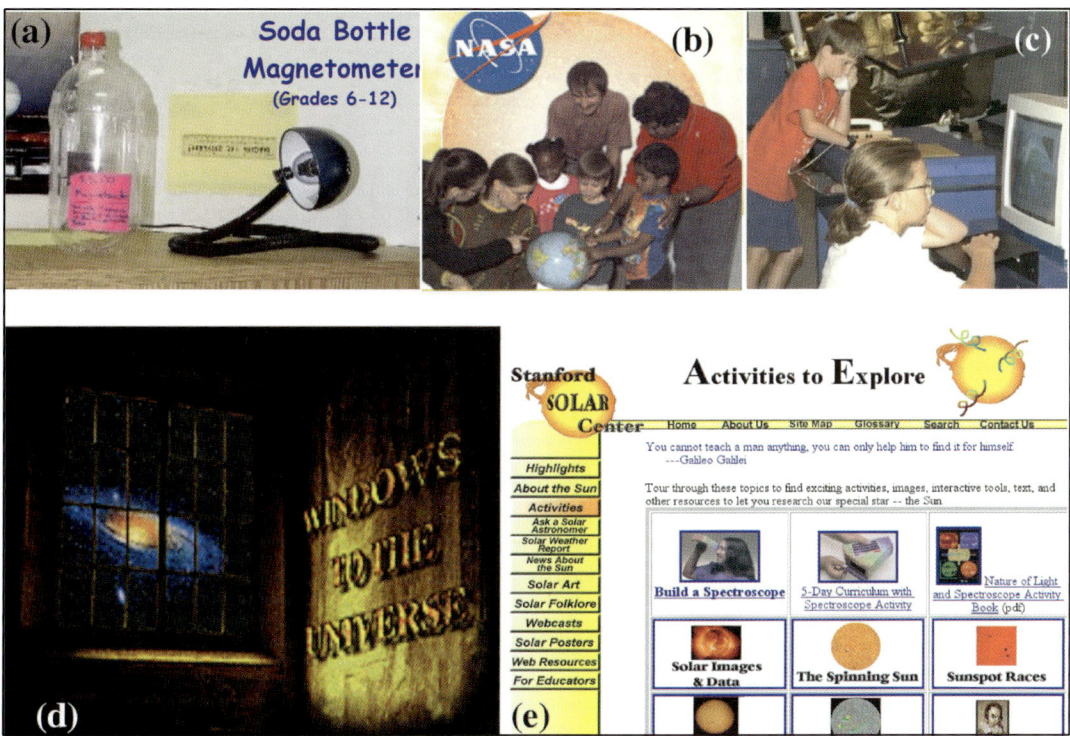

FIGURE 6.4 Solar and space physics enters the K-12 curriculum through physical science as well as Earth and space science topics. Scientists have partnered with education specialists to develop solar and space physics materials for K-12 students. Examples of such efforts include (a) information on how to build a simple device for measuring Earth's magnetic field, <http://image.gsfc.nasa.gov/poetry//workbook/magnet.html>; (b) the Sun-Earth Connection Education Forum, NASA's clearinghouse for connecting solar and space physics scientists and educators, <http://sunearth.gsfc.nasa.gov/index.htm>; (c) curriculum materials developed for formal and informal K-12 education by the Space Science Institute, <http://www.spacescience.org/Education/CurriculumDevelopment/1.html>; (d) *Windows to the Universe*, which delivers a wide range of resources in Earth and space science adapted for elementary, middle, and high-school levels, <http://www.windows.ucar.edu/>; and (e) the Stanford Solar Center, which provides activities for learning about solar physics in the classroom or at home, <http://solar-center.stanford.edu/activities.html>.

NATIONAL SCIENCE EDUCATION STANDARDS

Content Standards

Unifying concepts and processes
Science as inquiry
Physical science
Life science

Earth and space science
Science and technology
Science in personal and social perspectives
History and nature of science

Physical Science Standards

Levels K-4	Levels 5-8	Levels 9-12
Properties of objects and materials Position and motion of objects Light, heat, electricity, and magnetism	Properties and changes in properties of matter Motions and forces Transfer of energy	Structure of atoms Structure and properties of matter Chemical reactions Motions and forces Conservation of energy and increase in disorder Interactions of energy and matter

Earth and Space Science Standards

Levels K-4	Levels 5-8	Levels 9-12
Properties of Earth materials Objects in the sky Changes in Earth and sky	Structure of the Earth system Earth's history Earth in the solar system	Energy in the Earth system Geochemical cycles Origin and evolution of the Earth system Origin and evolution of the universe

with appropriate training, assisting in workshops that support teachers in implementing a significant science curriculum.[16]

Since the publication of the *Standards*, organizations such as the National Science Resources Center have produced guides for improving science education at different academic levels. For example, *Science for All Children*[17] discusses several exemplary school systems and suggests how parents, teachers, and administrators can go about improving the quality of science education in their schools. Many school systems have received grants from the U.S. Department of Education or the NSF (whose Office of Systemic Reform manages three large-scale reform projects: the Rural, Urban, and Statewide Systemic Initiatives) to improve instruction in science-related subjects.

Finding: Solar and space physics can play an important role in the implementation of national science education standards and state standards in science and technology, especially in middle and high schools in the subject areas of physical science and Earth and space science.

Education Initiatives at the NSF

Recent years have seen increasing interaction between the NSF's science organizations that support solar and space physics research—principally the Division of Atmospheric Sciences within the Directorate of Geosciences (GEO) and the Division of Astronomical Sciences within the Directorate of Mathematical and Physical Sciences—and programs in its Directorate of Education and Human Resources (EHR). Several valuable programs that enhance science education range from projects in local schools to national efforts carried out through EHR. For example, a 1997 report of the Geoscience Education Working Group recognized the important role of Earth scientists in education and recommended stronger ties and collaborations between GEO and EHR.[18] In response, Geosciences Education, a joint program initiated by GEO with EHR, was started in 1998. The program covers formal education from K-12 through graduate and postdoctoral training, as well as informal education carried out by about 20 competitively selected projects per year. These projects span the entire range of geophysics; few as yet include solar and space physics.

NSF support for solar and space physics educational activities could come about through the Opportunities for Enhancing Diversity in the Geosciences program in EHR's Division of Undergraduate Education[19] and in its Division of Elementary, Secondary, and Informal Education.[20] The NSF/

GEO diversity initiative presents a valuable opportunity to bring underrepresented groups into solar and space physics. However, the committee believes that NSF's education programs in the areas of solar and space physics could be expanded and strengthened. Collaborations between the Directorate of Education and Human Resources, the Directorate of Geosciences, and the Division of Astronomical Sciences should be designed to encourage solar and space physics researchers to bring exciting research results to the classroom and to the public.

Education and Public Outreach Initiatives at NASA

NASA's Office of Space Science (OSS) has established an extensive effort in education and public outreach. The OSS requires that approximately 2 percent of all mission costs be spent on education and public outreach (see Figures 6.1 and 6.4 above). Most of the activities supported build on existing programs at NASA centers and in museums, school systems, or educational institutions. Many are highly leveraged, with the aim of having a substantial national impact.

A recent review and assessment of OSS's education and public outreach program gave high marks for the program's information content but noted that the limited financial resources of most schools pose a challenge to getting space science into the classroom.[21] The assessment further noted that difficulties can arise at times in such programs, because scientists and non-college-level educators come from different professional cultures and often approach a project from incompatible viewpoints.

In addition to supporting education and public outreach programs that are linked to specific flight missions, NASA's OSS provides supplements in the form of small grants that support participation by individual scientists. The scope of these supplements has recently been expanded to allow, for example, budgets for education and outreach of up to 5 percent of the primary research grant, as well as bundling of institutions' education and outreach activities into larger grants. These supplements to grants have enabled solar and space physics researchers to become involved in education and public outreach activities at the local level.[22]

Finding: NASA-funded education and public outreach projects encourage and permit researchers to collaborate with educators on a wide variety of educational activities related to solar and space physics, many of which have a substantial impact on public awareness of issues in solar and space physics and their link to broader science and technology concerns.

NOTES

1. "The harsh fact is that the U.S. need for the highest quality human capital in science, mathematics, and engineering is not being met," according to the Hart-Rudman Commission on National Security for the 21st Century, which makes a series of recommendations to improve the government's ability to address the national security challenges of the new century. Several major recommendations from the commission deal with the nation's scientific research and education enterprises (*FYI: The American Institute of Physics Bulletin of Science Policy News,* No. 23, March 1, 2001).

2. "It is in college where future scientists and college faculty are recruited and prepared for graduate study; where our nation's elementary and secondary teachers, educators of America's youth, are equipped; and where tomorrow's leaders gain the background with which to make critical decisions in a world permeated by vital issues of science and technology." See report from the NSF-supported Project Kaleidoscope, "What Works: Building Natural Science Communities, Vol. I," 1991, available online at <http://www.pkal.org/template2.cfm?c_id=394>.

3. Nicholson, S., and P. Mulvey, *American Institute of Physics Report R-394-7,* September 2001.

4. Concerns about the loss of solar research faculty at research institutions such as the California Institute of Technology, Stanford University, the University of Maryland, and the University of Colorado were expressed in a report of the National Research Council (NRC, *Ground-Based Solar Research: An Assessment and Strategy for the Future,* National Academy Press, Washington, D.C., 1998, pp. 4-5, available online at <http://www.nap.edu/catalog/9462.html>). While noting that institutions such as the New Jersey Institute of Technology and Montana State University had hired new faculty, the report expressed concern about the likely effect of a general loss of faculty on training for the next generation of graduate students, as well as about the state of university-based instrumentation programs, which are widely seen as essential to future instrument development and to the training of new researchers with hands-on experience. Finally, the report noted that "existing programs are few in number and rely on precarious grant-based funding" (p. 5). Overall, 3 percent of physics faculty members are retiring every year, and this rate is expected to increase (Ivie, R., K. Stowe, and R. Czujko, *American Institute of Physics Report R-392-4,* March 2001).

5. Asynchronous learning, also called location-independent learning, occurs when students learn the same material at different times and in different locations. The asynchronous learning environment provides students with teaching materials and tools for registration, instruction, and discussion. Asynchronous learning requires the ability to maintain communication without having to meet at the same place at the same time. Students attending a lecture or laboratory session together are participating in synchronous learning. See <http://www.engin.umich.edu/~cre/asyLearn/index.html>.

6. Richardson, John T.E., *Researching Student Learning,* Open University Press, 2000.

7. *Physics Today,* April 2001, p. 47.

8. ". . . many studies have shown that the undergraduate programs most successful at producing scientists are those that include research and publication in refereed journals. . . . Students who have the opportunity for research complete their science programs in greater numbers than those who do not" (Gavin, Robert, The role of research at undergraduate institutions: Why is it necessary to defend it?, *Academic Excellence,* M.P. Doyle, ed., Research Corporation, Tucson, Ariz., 2000).

9. Seymour, E., and N.M. Hewitt, *Talking About Leaving: Why Undergraduates Leave the Sciences,* Westview Press, 1997.

10. Tobias, S., *They're Not Dumb, They're Different: Stalking the Second Tier,* Research Corporation, Tucson, Ariz., 1990.

11. NRC, *Transforming Undergraduate Education in Science, Mathematics, Engineering, and Technology*, National Academy Press, Washington, D.C., 1999.

12. Mazur, E., *Peer Instruction*, Prentice-Hall, Englewood Cliffs, N.J., 1997; McDurmott, L.C., 1990: What we teach and what is learned—Closing the gap, *American Journal of Physics* 59, pp. 301-315, 1991; Redish, E., Implications of cognitive studies for teaching physics, *American Journal of Physics* 62, pp. 796-803, 1994; Redish, E., and N. Steinberg, Teaching physics: Figuring out what works, *Physics Today*, January 1999; and Mestre, J.P., Learning and instruction in pre-college physical science, *Physics Today*, September 1991.

13. National Science Teachers Association, *College Pathways to the Science Education Standards*, 2001; NRC, *Transforming Undergraduate Education in Science, Mathematics, Engineering and Technology*, National Academy Press, Washington, D.C., 1999; NRC, *Educating Teachers of Science, Mathematics, and Technology: New Practices for the New Millennium*, National Academy Press, Washington, D.C., 2000; National Commission on Mathematics and Science Teaching for the 21st Century (the Glenn Commission), *Before It's Too Late*, U.S. Department of Education, Washington, D.C., 2000; "Prospective and practicing teachers must take science courses in which they learn science through inquiry, having the same opportunities as their students will have to develop understanding" (NRC, *National Science Education Standards*, National Academy Press, Washington, D.C., 1996, p. 60).

14. NRC, *Enhancing Undergraduate Learning with Information Technology*, National Academy Press, Washington, D.C., 2002.

15. NRC, *National Science Education Standards*, National Academy Press, Washington, D.C., 1996.

16. NRC, *Revolution in Earth and Space Science Education*, National Academy Press, Washington, D.C., 2002.

17. NRC, *Science for All Children: A Guide to Improving Elementary Science Education in Your School District*, National Academy Press, Washington, D.C., 1997.

18. NSF, *Geosciences Education: A Recommended Strategy*, Washington, D.C., 1997.

19. For example, programs for advanced technological education and for course, curriculum, and laboratory improvement and the computer science, engineering, and mathematics scholarship program.

20. For example, the programs for informal science education, instructional materials development, and teacher enhancement.

21. An independent critique of the whole of the OSS's education and public outreach program was made by S.B. Cohen and J. Gutbezahl from the Program Evaluation and Research Group, Lesley University. See S.B. Cohen, J. Gutbezahl, and J. Griffith, *Office of Space Science Education/Public Outreach: Interim Evaluation Report, October 2001-June 2002*, NASA, October 2002.

22. The 2001 report of OSS's education and public outreach activities suggests 15 or so such solar and space physics activities out of about 60 total projects.

7
Strengthening the Solar and Space Physics Research Enterprise

The prospects for substantial advances in solar and space physics depend on a significantly improved understanding of the key physical processes that are encountered in space. As emphasized in previous chapters of this report, achieving these advances will require strengthening the national infrastructure for solar and space physics research in a number of areas. The committee has identified several areas in particular in which effective program management and the appropriate policy actions could enhance the ability of the solar and space physics communities to address the science challenges presented in Chapter 1: development of a stronger research community, cost-effective use of existing resources, ensuring cost-effective and reliable access to space, improving interagency cooperation and coordination, and facilitating international partnerships. The following sections describe each of these areas and offer recommendations for optimizing the science return of solar and space physics over the next decade.[1]

A STRENGTHENED RESEARCH COMMUNITY

For decades before the first scientific satellites flew in 1957 and 1958, studies of geomagnetism, the aurora, cosmic rays, and related topics had been carried out at universities, in some government laboratories, and in the few industrial organizations that needed the information for their businesses. Since the advent of the space age, ground-based research in solar and space physics has been conducted principally in the universities with support from the National Science Foundation and the Air Force. Agencies such as NOAA, DOE, and the Department of the Interior have performed some research related to their governmental missions, but their support of university (or other outside) researchers has not been of any great significance. There was relatively little nongovernmental expertise in spaceflight capabilities at the time NASA was established.[2] However, very early in its history NASA strongly encouraged university participation by offering incentives in the form of financing for buildings, the sponsorship of students, and the creation of specialized centers of expertise.

The successful pursuit of solar and space physics depends on an effective and collaborative working balance among universities, government laboratories, industry, and not-for-profit research organizations, drawing on the unique contributions that each can make. The university is the means for recruiting and training the young people who will carry on the next generation of research activities in academic, public, and private sector laboratories. In particular, universities introduce young people to the science of Earth and the solar system and provide them with a science background for the remainder of their lives, independent of their educational degree or life work. Universities subsequently educate some of the students at a graduate level through both classroom instruction and working experience in ongoing research projects. These projects can involve scientific instrument development and deployment, data analysis, or theoretical calculations and interpretation. The universities also provide a base of operations for the science faculty who are responsible for the education and training of students and offer a stimulating intellectual environment of scientific cooperation and individual invention.

Strengthening University Involvement in Spaceflight Programs

The active involvement of university faculty and students in the nation's spaceflight program is seen today as one of the major strengths of U.S. space research. Indeed, the level of university involvement in this country is almost unique in the world when compared with the space programs of other nations. At the same time, an environment of what can best be termed "creative tension" has long existed between the NASA centers and university researchers when it comes to opportunities for funding and spaceflight experiments.

A number of factors contribute to the committee's serious concern regarding the future of universities as sites for NASA-supported solar and space physics research. Many of these factors also affect other disciplines involved in space-based experiments. For example, NASA's episodic funding of individual university research programs leads to gaps between major grants when technical staff might have to be dismissed for lack of funds. Such disruptions could seriously hamper the ability of a university research group to develop new concepts and instrumentation so as to remain competitive for future opportunities. In particular, universities seem to be unable to remain as stable as NASA centers. A discouraging situation can arise when federal solicitations of proposals for instruments for space missions are subsequently canceled, as occurred most recently in the case of the Solar Probe mission. In such cases, the preparation of a well-thought-out competitive proposal (often involving industry and/or a nonprofit organization as a partner) can be an immense and costly effort. The subsequent decision to cancel can have a lasting negative effect on scientific productivity, not just in the affected university research group but ultimately in the nation itself.

Recommendation: NASA should undertake an independent outside review of its existing policies and approaches regarding the support of solar and space physics research in academic institutions, with the objective of enabling the nation's colleges and universities to be stronger contributors to this research field.

This review would look at universities as research sites[3] that contribute significantly to the nation's solar and space physics program. It would examine in depth such topics as ways of sustaining meaningful partnerships with NASA laboratories and centers, the competitive process for space hardware procurement to maximize opportunities for university participation, and methods for ensuring reasonable stability for critical-mass technical teams in university research groups and laboratories.

Strengthening University Participation in National Research Facilities

National facilities play an essential role in the nation's science and engineering efforts, but only when they are used effectively. The committee believes that national facilities should be operated as more than national research institutes—they should be widely available to outside users. The scientific effectiveness of some national facilities, e.g., the National Solar Observatory (NSO) and the large radar and optical observatories of the NSF's Upper Atmosphere Facilities Program, is compromised by the limited funding available to outside investigators. The NSO budget contains an insignificant amount of funding (less than 5 percent of the total) for university users of its facilities. Unlike that of the Space Telescope Science Institute, for example, NSO telescope use by outside investigators is not leveraged for maximum return on the nation's investment in these instruments. The NSF's radar and optical observatories are important to a different segment of the solar and space physics research community. A wide range of users access these facilities and their data as guest researchers and via the Internet. Although provisions are in place to accommodate visitors and their instruments and to process and provide data for them to use in their research, no budgeting line has been established to support the research activities of the outside users of these facilities.

Recommendation: NSF-funded national facilities for solar and space physics research should have resources allocated so that the facilities can be made widely available to outside users.[4]

Such funding would allow for substantial peer-reviewed guest investigator programs and for substantive community involvement in the definition, design, oversight, and development of new facilities such as the Advanced Technology Solar Telescope, which is currently being developed under the leadership of NSO.

COST-EFFECTIVE USE OF EXISTING RESOURCES

Return on investment is optimized not only through the judicious funding and management of new observing systems, but also through the maintenance, upgrading, funding, and management of existing systems. Facilities for ground- and space-based solar-terrestrial research that have already been developed and paid for are currently operating and returning data. It is often feasible and cost-effective to employ as many such assets as possible in a research program.

Ground-based systems include arrays of passive sensing instruments such as riometers, magnetometers, cosmic ray sensors, and optical systems, as well as large facility-class installations such as ionosphere-sounding radars. Ground-based assets (most of which are supported by the NSF) can often be improved at relatively low cost to become part of a new research program—both the sensors and the data collection devices of the instrumentation can be upgraded, and the locations and distributions of instrument arrays can be changed.

For most space assets, it is not practical or even possible to change the instrumentation. However, many existing solar and space physics missions still return essential data at relatively low cost. The two Voyager spacecraft at distances greater than 60 AU, the Ulysses mission out of the ecliptic over the solar poles, the ACE and Wind missions upstream of Earth at approximately 0.01 AU (the L1 point), and the IMP-8 mission nearer Earth remain important sources of solar and space physics data. Solar missions in this category are SOHO (at the L1 point) and TRACE (in Earth orbit). Magnetospheric missions that continue to be very productive include Polar, IMAGE, the Japanese Geotail mission, and the European Cluster mission.

NASA considers the continuing operation of existing space assets to be mission extensions, even if the data acquired by them are used for new research thrusts. These extended missions often suffer in resource allocation reviews despite the modesty of their budget requirements in comparison with those of new flight missions. To be sure, it is critical to fly new missions and generate new discoveries; however, as the research field moves into an era that involves the development and validation of prediction tools, the maintenance of the existing fleet of missions (especially those that return data from unique locations in the heliosphere and those that are important to space weather) should become more important.

Many existing assets, both ground- and space-based, can contribute significantly to addressing the scientific challenges set forth in this report.

Recommendation: The NSF and NASA should give all possible consideration to capitalizing on existing ground- and space-based assets as the goals of new research programs are defined.

It might be that management and programmatic changes to these facilities or missions—perhaps, for example, the transfer of operations and data acquisition to academic or other organizations—would result in considerable cost savings. The exploration of such possibilities is strongly encouraged as the decadal research program is developed. Further, it might be that some of the assets could be used for operational purposes, and this should be considered as well. For example, in Chapter 5 the committee addresses the continued acquisition of solar wind data at the L1 point and recommends, in view of the importance of such measurements for space weather operations, that NOAA become the responsible entity.

ACCESS TO SPACE

The continuing vitality of the nation's space research program is strongly dependent on having cost-effective, reliable, and ready access to space that meets the requirements of a broad spectrum of diverse solar and space physics missions. The solar and space physics research community is especially dependent on the availability of a wide range of suborbital and orbital flight capabilities to carry out cutting-edge science programs, to validate new instruments, and to train new scientists. Difficulties in one or more of these program elements can translate into fewer (or no) research opportunities and increased mission costs. Several programmatic issues relating to access to space are discussed below.

Suborbital Program

The platforms for the Suborbital Program are sounding rockets (see Figure 7.1), high-altitude balloons, and aircraft.

• *Sounding rockets.* Sounding rockets are critical for the investigation of important small-scale processes in the terrestrial ionosphere, generally from altitudes between about 90 km and several hundred kilometers. Rockets are used to fly stand-alone individual payloads for targeted space plasma research, often in close collaboration with orbital and ground-based measurements. Besides addressing frontier space plasma problems such as small-scale particle acceleration regions, sounding rocket investigations have also served as exemplary tools for the development of scientific ideas and measurement technologies, and they have had a significant level of student participation, often far out of proportion to the program costs. The often fast turnaround from scientific concept through engineering of the instrumentation, flight, and data return and analysis (such speed is characteristic also of balloons and airplane platforms; see below) is entirely consistent with the educational objectives of universities.

In recent years, for a variety of reasons that appear to have included program management and resource allocation decisions, the number of rocket flight opportunities has been decreasing. Illustratively, in FY 2001 fewer than half as many NASA sounding rockets were launched as in the 1980s and 1990s, when there were, on average, 25 launches every year. This decrease in flight opportunities does not appear to have been based on any comprehensive assessment of the program's scientific merits or its opportunities or on peer-reviewed determinations of the adequate size of the program. Rather, some resources were moved to the now discontinued UNEX small satellite program, and management changes for the rocket program did not result in increased science flight opportunities.[5]

• *Balloons.* High-altitude balloons with a maximum payload of 1,500 kg are capable of reaching altitudes that are above 99.5 percent of the atmosphere. A new super-pressure balloon is being tested that will provide 100-day flights. Balloons contribute to research in cosmic ray physics, solar physics, and atmospheric chemistry and physics. Balloons, like sounding rockets, also have platform features that are ideal for university-based education programs.

FIGURE 7.1 A Black Brant XII sounding rocket carrying aloft an experiment to measure atomic oxygen emissions in Earth's upper atmosphere. Sounding rockets are used for a variety of important research objectives in solar and space physics, from in situ study of ion acceleration processes in the high-latitude ionosphere to remote-sensing observations of the Sun's corona. Because they are relatively inexpensive and require less time to develop and implement than satellite investigations, sounding rocket experiments have proven to be a valuable education asset for training as well as an important research tool. Courtesy of P.J. Eberspeaker (NASA Wallops Flight Facility).

- *Aircraft.* The NASA research aircraft program encompasses a number of different aircraft, most of them based at the NASA Ames and Wallops facilities. Such platforms are of greatest importance for those problems in space physics that address key issues in atmosphere-ionosphere coupling. The Stratospheric Observatory for Infrared Astronomy (SOFIA) telescope, currently under development on a modified Boeing 747, will provide opportunities for studies of planetary atmospheres, especially—in the context of space plasma physics—of planetary magnetospheric and aeronomic processes such as aurora production.

Finding: Suborbital flight opportunities are very important for advancing numerous key aspects of solar and space physics research and for their significant contributions to education.

Recommendation: NASA should revitalize the Suborbital Program to bring flight opportunities back to previous levels.

Revitalizing the Suborbital Program will be necessary in order to accomplish the new science directions identified in this survey (see the priorities in Chapter 2). Implementing this recommendation will probably require that NASA, in collaboration with the solar and space physics research community, perform a thorough, in-depth, and independent review of the programmatic aspects of the Suborbital Program, identifying its strengths and weaknesses.

Orbital Program

The diversity of flight missions needed in the future calls for a broad range of launch vehicles. Currently there is a limited choice of launch vehicles available for space physics missions, and launch costs can be a large proportion of overall mission costs. A recent report from the Space Studies Board[6] discussed many of the launch vehicle challenges that face solar and space physics research.

For small payloads of the Small Explorer (SMEX) category (cost cap of $80 million, including launcher), the only available vehicle is the airplane-launched Pegasus rocket. The cost of a Pegasus launch can consume nearly 25 percent of the allotted cost, thus limiting the size of the payload and the spacecraft, as well as the resources available for mission operations and data analysis.

The Taurus and the Delta II vehicles are used for Medium-Class Explorer (MIDEX) solar and space physics missions (cost cap of $185 million, including launcher). A Delta II launch can account for as much as 33 percent of the total cost allowed for the mission. While the Taurus is considerably cheaper, its reduced performance in comparison with a Delta II means that solar and space physics experiments flown on Taurus vehicles are confined to those that can be carried out from low-Earth orbit.

The paucity of small and inexpensive boosters in NASA's launch portfolio will probably cause more missions than ever (not all of which are scientific) to be piggybacked on more expensive spacecraft on even larger boosters. Launches in the Air Force's Space Test Program can employ an adapter ring capable of carrying and deploying several small satellites, but the use of such a ring for civilian science satellites has not been generally possible. Even when piggybacking is programmatically possible, it might force a schedule or level of quality that is incommensurate with the optimal timing of the mission or with its class.

One way of facilitating access to space would be to allow U.S. payloads to be launched on foreign vehicles. For example, the European Ariane launcher routinely launches small secondary payloads at a modest cost ($1 million per 100-kg payload) with a simple, standard interface. U.S. launch policies currently prohibit U.S. researchers from exercising such an option. This prohibition stems largely from a desire to encourage the use of U.S. launch vehicles.

Low-cost launch vehicles with a wide spectrum of capabilities are critically important for the next generation of solar and space physics research as delineated in this survey.

Recommendations:

1. NASA should aggressively support the engineering research and development of a range of low-cost vehicles capable of launching payloads for scientific research.

2. NASA should develop a memorandum of understanding with DOD that would delineate a formal procedure for identifying in advance flights of opportunity for civilian spacecraft as secondary payloads on certain Air Force missions.

3. NASA should explore the feasibility of similar piggybacking on appropriate foreign scientific launches.

The Defense Advanced Research Projects Agency (DARPA) is working to develop a very low cost launcher to space for small payloads—the Responsive Access Small Cargo and Affordable Launch (RASCAL) program. If it is realized, the program's objective (to launch a 75-kg payload to low-

Earth orbit at a cost of $750,000) could benefit several space plasma physics programs.

Comparative Planetary Plasma Physics and the Discovery Program

The comparative study of planetary ionospheres and magnetospheres is a central theme of solar and space physics research (see Figure 7.2).[7] Further, the planetary environments of the solar system are testbeds to validate models of how solar effects propagate through the heliosphere and how they interact with atmospheres and magnetospheres. In the early days of solar system exploration, large missions such as Pioneer, Voyager, Galileo, and Cassini could accommodate planetary geology, atmospheric science, and space physics payloads. Those days of occasional, large, complex spacecraft have been followed by the budget-constrained missions of the Discovery program. Such missions are so limited in terms of cost, mass, power, and data rate that they are generally not able to address both planetary and space physics objectives. A solution to this dilemma is to open the Discovery competition to missions that exclusively address planetary space physics objectives.

Recommendation: The scientific objectives of the NASA Discovery program should be expanded to include those frontier space plasma physics research subjects that cannot be accommodated by other spacecraft opportunities.

Controlling Spaceflight Mission Cost Growth

The use of cost caps during much of the 1990s, together with the placement of responsibilities for mission development and success in the hands of a mission principal investigator (PI), played a significant role in many highly successful solar and space physics missions, including the Solar Anomalous and Magnetospheric Particle Explorer (SAMPEX), Fast Auroral Snapshot Explorer (FAST), TRACE, ACE, and IMAGE. Besides being very successful scientifically, all of these solar-terrestrial missions were developed at a cost less than their allocated budgets. The PI model that was used for these Explorer missions was highly successful by any standard. Strategic missions such as those under consideration for the Solar Terrestrial Probes and Living With a Star mission lines could benefit from emulating some of the management approach and structure of the Explorer missions.

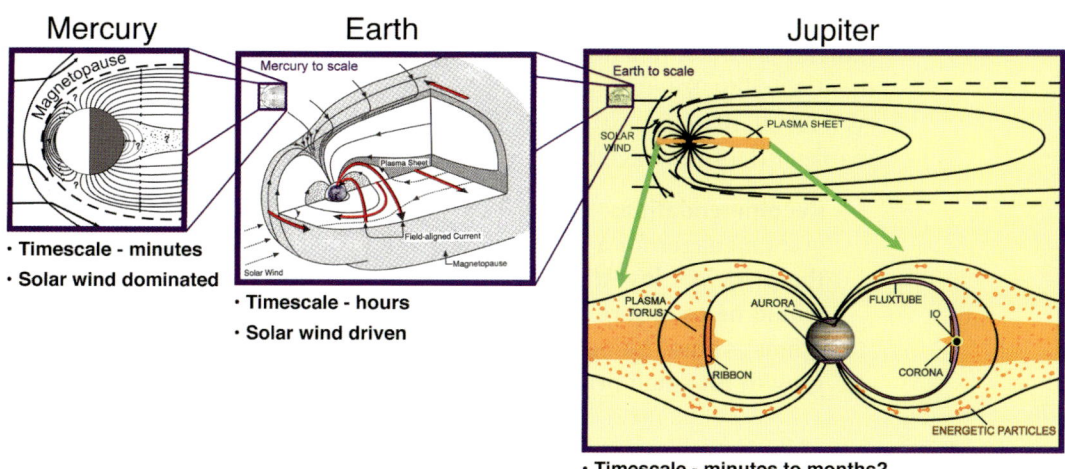

FIGURE 7.2 Planetary magnetospheres show great diversity in size, structure, composition, and dynamics. This diversity reflects differences in the strength and orientation of the planetary magnetic fields, the sources of magnetospheric plasma, and the relative roles of planetary rotation and the solar wind in powering the magnetosphere. A major theme of space physics is to understand the similarities and differences among the various magnetospheres of the solar system and to test our understanding of fundamental plasma physical processes by observing how they work in different magnetospheric environments. Courtesy of F. Bagenal (University of Colorado).

Many of the major science objectives of solar and space physics research are naturally suited for implementation by a PI.

Cost caps can be effective in controlling mission cost growth. However, caps will not work if they are not taken seriously or not enforced, or if costs are beyond the control of the developer or the PI. Experience has shown that a successful cost-capped system requires that the rules for development be thoroughly understood by the developer (usually the PI) before development begins and, further, that the rules should not be changed later on.

Recommendation: NASA should (1) place as much responsibility as possible in the hands of the principal investigator, (2) define the mission rules clearly at the beginning, and (3) establish levels of responsibility and mission rules within NASA that are tailored to the particular mission and to its scope and complexity.

Unfortunately, such tailoring often proved difficult in the past because individual NASA functional organizations (such as Earned Value Management, Quality, Safety, and Verification) tend to impose nonnegotiable rules. As a result, the principal NASA official interacting with a mission PI and/or manager does not have the authority to negotiate all aspects of the project.

Recommendation: The NASA official who is designated as the program manager for a given project should be the sole NASA contact for the principal investigator. One important task of the NASA official would be to ensure that rules applicable to large-scale, complex programs are not being inappropriately applied, thereby producing cost growth for small programs.

INTERAGENCY COOPERATION AND COORDINATION

Over the years interagency coordination has often yielded greater science returns than have single-agency activities. For example, NSF, NASA, and ONR have coordinated their scientific investigations from high-altitude balloons. The International Magnetosphere Study in the 1970s was an excellent example of an interagency (NSF, NASA, NOAA) cooperative program that extended to the international scene as well. NASA's Thermosphere Ionosphere Mesosphere Energetics and Dynamics (TIMED) mission and NSF's CEDAR initiative have been coordinated in a manner that collectively returns more new knowledge than if the two programs had been run entirely separately. NASA and DOD are collaborating in missions of opportunity to provide data that would not otherwise be available. More recently, the development of the NPOESS spacecraft marked the beginning of a coordination among three agencies, NOAA, DOD and NASA, that will be of value to a wide variety of scientific and operational activities in space. In contrast, depending on the final schedules, the amount of data overlap between NASA's planned Solar Dynamics Observatory (SDO) (launch in 2007 with a design life of 5 years) and NSF's Advanced Technology Solar Telescope (ATST) (not anticipated to be completed until 2010) may not be sufficiently great. It will be necessary, therefore, as ATST comes on line to evaluate the measurement overlap between the two facilities and to determine the science and resource requirements for maintaining SDO measurements in order to obtain the desired overlap in data coverage.

In the future, a research initiative within one agency could trigger a window of opportunity for a research initiative in another agency. Such an eventuality would leverage the resources contributed by each agency.[8]

Recommendation: The principal agencies involved in solar and space physics research—NASA, NSF, NOAA, and DOD—should devise and implement a management process that will ensure a high level of coordination in the field and that will disseminate the results of such a coordinated effort—including data, research opportunities, and related matters—widely and frequently to the research community.

Recommendation: For space-weather-related applications, increased attention should be devoted to coordinating NASA, NOAA, NSF, and DOD research findings, models, and instrumentation so that new developments can quickly be incorporated into the operational and applications programs of NOAA and DOD.

FACILITATING INTERNATIONAL PARTNERSHIPS

International Cooperation and Collaboration

The geophysical sciences, and in particular solar and space physics, address questions of global scope and inevitably require international participation for their success. This is particularly the case for ground-based solar and space physics research. For example, collaborative research with other nations allows the United States to obtain data from other geographical regions that are necessary to determine the global distributions of space processes. Studies in space weather cannot be successful without strong participation from colleagues in other countries and their research capabilities and assets, in space and on the ground.

Even if financial considerations were irrelevant, the United States would be compelled to join forces with the international community to achieve the full scientific return from its own investments in space research. Such considerations are not, however, irrelevant. By working with other nations on joint programs such as the International Solar-Terrestrial Physics program, sharing the burden of instrumentation for moderate (e.g., STEREO) and large (e.g., Ulysses, SOHO, and Cassini) space projects, and collaborating in Antarctic programs and in ground-based facilities such as GONG and the SuperDARN[9] radars, the U.S. solar and space physics research community has executed an ambitious and effective research program in a cost-effective manner.[10]

Finding: The United States has greatly benefited from international collaborations and cooperative research in solar and space physics.

The benefits of these international activities have allowed the implementation of programs that would not otherwise have been possible and have permitted the acquisition of data and understanding that are essential for the advancement of science and applications.

International Traffic in Arms Regulations

Much of the ease with which international cooperation in space-based research was achieved in the past has been lost in the last several years as regulatory changes intended to apply to arms and related matters have been applied to scientific activities. In FY 2000, responsibility for satellite technology export licensing, regulated under the International Traffic in Arms Regulations (ITAR), was transferred from the Department of Commerce to the Department of State.[11] Before the transfer of responsibility, scientific satellites had routinely been granted an exclusion from the application of these regulations. Although a directive excluding scientific satellites from the regulations remains in effect, ambiguities in the statements of requirements have led some federal and research institutions to erect barriers to the exchange of scientific information and instrumentation that could, in a restrictive interpretation, fall under ITAR.[12]

The impact of the uncertainties related to the application of ITAR to scientific research has been a subject of intense discussion in the research community and among federal legislators and affected federal agencies. This is an important issue since international collaboration in space science research has been encouraged and fostered as a matter of national policy since the early days of the civil space program. In recent years, the situation as it affects the research community has continued to evolve. In March 2002 an interim rule was issued by the State Department that relaxes and clarifies some of the regulations as they apply to university-based research (*Federal Register*, Vol. 67, No. 61, pp. 15099-15101, March 29, 2002). ITAR licenses will not be required for the export of scientific satellite hardware or information from U.S. universities to members of NATO and major non-NATO allies. The restrictions that remain on citizens from countries not considered principal allies of the United States will still affect university research, and many see the separation of students into two categories as unworkable. Furthermore, if the hardware is built as a part of a collaborative effort between universities and industry, as is normal these days, the ITAR conditions again become operative. In summary, while the 2002 changes in the rules are viewed by export officials as easing life for university researchers, and the research community agrees, formidable issues continue to affect university researchers and their scientific colleagues and collaborators in industry and nonprofit organizations. These remain to be worked on in the months ahead.

The committee recognizes the continuing critical national imperative for international arms control as well as the long-standing national interest in international scientific space research. It is in this context and with the aim of expediting international collaborations that involve scientific data, instrument characteristics, and instrument exchanges that the following recommendation is made.

Recommendation: Because of the importance of international collaboration in solar and space physics research, the federal government, especially the State Department and NASA, should implement clearly defined procedures regarding exchanges of scientific data or information on instrument characteristics that will facilitate the participation of researchers from universities, private companies, and nonprofit organizations in space research projects having an international component.

NOTES

1. The individual panel reports of this survey (*The Sun to the Earth—and Beyond: Panel Reports,* in preparation) also address these and related structural issues.

2. Newell, H., *Beyond the Atmosphere: Early Years of Space Science,* NASA, Washington, D.C., 1980; NRC, *A Review of Space Research: The Report of the Summer Study Conducted Under the Auspices of the Space Science Board of the National Academy of Sciences at the State University of Iowa, Iowa City,* National Academy of Sciences, 1962.

3. The broader role of the universities in solar and space physics education is addressed in Chapter 6.

4. The recent decadal strategy survey for astronomy and astrophysics made a similar recommendation regarding NSF facilities for astronomy (NRC, *Astronomy and Astrophysics in the New Millennium,* National Academy Press, Washington, D.C., 2001, p. 5).

5. For example, in a June 2000 letter to the Office of Space Science, the NASA Sounding Rocket Working Group pointed out the financial crisis the Sounding Rocket Program was experiencing at the Wallops Flight Facility. Under the heading "The Current Problem," the letter stated as follows:

> The Sounding Rocket Program supported a flight rate of approximately 25-30 rockets each year with budget of $30.6M/year through FY95. Subsequently, $6M in FY 96 and an additional $1M in FY97 were diverted to the UNEX program, leaving an operating budget of $23.6M/year since FY97. During this same period, the program was directed to privatize its operations, with the NASA Sounding Rocket Operations Contract (NASROC) taking over almost all of the traditional projected management, technical, and implementation duties at Wallops in mid-1999 . . . . Although the science community is pleased with the NASROC technical performance to data [sic], no significant cost savings have been realized in the brief 1.3 years of the NASROC contract nor are significant [savings] anticipated in the foreseeable future. The program has

> survived during the last few years by using reserve funds as well as working off of its inventory. . . . As the reserve funds are now depleted, NASA/GSFC management has requested Overguide Funds of approximately $10M for the Program in FY01 in its most recent POP request, with a similar figure for the years FY02 and beyond. . . .

Under the heading "Implications," the letter stated as follows:

> Wallops and NASROC have concluded that without an augmentation they will only be able to handle 9 missions a year, which will cause the delay of 46+ missions now being planned, designed, built, and/or tested for launch in FY01 and FY02. In addition to the loss of scientific research opportunities, a much more serious consequence will result: by only launching 9 missions a year, the core competency of the program will be lost. Indeed, NSROC management informs us that they will begin to lay off 20% of their work force this October as a result of inadequate funds to cover privatization and the loss of the civil servant workforce. Without this critical know-how, the success rate for the few rockets that will be flown will be in jeopardy. Furthermore, each of these now even more precious rockets becomes much more expensive since the program will lose its advantage of economies of scale.

6. National Research Council, *Assessment of Mission Size Trade-Offs for NASA's Earth and Space Science Missions*, National Academy Press, Washington, D.C., 2000.

7. See National Research Council, *The Sun to the Earth—and Beyond: Panel Reports*, Panel on Atmosphere-Ionosphere-Magnetosphere Interactions, The National Academies Press, Washington, D.C., 2003, in press.

8. A recent report from the NRC's Committee on the Organization and Management of Research in Astronomy and Astrophysics (*U.S. Astronomy and Astrophysics: Managing an Integrated Program*, National Academy Press, Washington, D.C., 2001) discusses research coordination among agencies in astronomy and makes several cogent recommendations related to this field.

9. The Super Dual Auroral Radar Network (SuperDARN) is a ground-based network of high-frequency radars that are used to study Earth's ionosphere.

10. Recent examinations of a number of facets of international cooperation in space research are contained in reports from the Space Studies Board of the National Research Council: National Research Council and European Science Foundation, *U.S.-European Collaboration in Space Science*, National Academy Press, Washington, D.C., 1998; Science Council of Japan, European Science Foundation, and National Research Council, *U.S.-European-Japanese Workshop on Space Cooperation: Summary Report*, National Academy Press, Washington, D.C., 1999.

11. The complete texts of ITAR regulations can be found at <http://www.pmdtc.org/>. This site also has links to recent (April 29, 2002) amendments to ITAR.

12. See U.S. Congress, House Committee on Appropriations, *Departments of Veterans Affairs and Housing and Urban Development and Independent Agencies Appropriations Bill, 2002, Report to Accompany H.R. 2620*, 107th Cong., 1st sess., 2001, H. Rept. 107-159. References to ITAR in the report include the following (p. 85):

> As mentioned in the conference report accompanying the fiscal year 2001 appropriations bill, Public Law 105 261 transferred responsibility for satellite technology export licensing from the Department of Commerce to the Department of State to be regulated under the International Traffic in Arms Regulations (ITAR). While scientific satellites are still covered by the fundamental research exclusion provided by National Security Directive 189, the unfortunate and unintended consequence of the jurisdictional move has been that university-based fundamental science and engineering research, widely disseminated and unclassified, has become subject to overly restrictive and inconsistent ITAR direction.

Appendixes

A

Statement of Task

Background: The last integrated strategy for solar and space physics was released by the NRC in 1995. Since that time, there have been dramatic scientific developments and a significant evolution in relevant federal programs. In the space arena these developments stem from the launches and successful operation of the Wind, Geotail, SOHO, Polar, FAST, ACE, TRACE, IMAGE, and Cluster-II missions. These missions have helped revolutionize solar physics, provide a new level of understanding of important processes in space plasma physics, and create a new basis for characterizing and predicting space weather. Over the same period, the relevant federal agencies have taken steps to build on the new level of scientific progress by embarking on new efforts such as the National Space Weather Program, the Relocatable Radar (formerly the Polar Cap Observatory), and Living with a Star. Furthermore, the NSF Geospace Environment Modeling (GEM) program has initiated its second and third campaigns, the international Super Dual Auroral Radar Network (SuperDARN) has established effective collaboration among a large number of high-frequency radar programs; and the community-wide Solar, Heliospheric, and Interplanetary Environment (SHINE) initiative has spawned a number of important activities related to the National Space Weather Program. As a consequence of all these developments, the preparation of a comprehensive scientific assessment and strategy for the field of solar and space physics that looks across the interests of all agencies, both ground- and space-based, is especially timely.

Plan: The study will be organized in a manner similar to the 'decadal survey' that is regularly conducted by the astronomy and astrophysics community. The Committee on Solar and Space Physics will establish a 14-person survey committee to carry out the study with input from five panels, each of which will have approximately 10 members. Most Committee on Solar and Space Physics members will serve either on the survey committee or the panels, with additional membership drawn from the relevant research communities.

The study will generate consensus recommendations from the solar and space physics community regarding a systems approach to theoretical, ground-based, and space-based research that encompasses the flight programs and focused campaigns of NASA, the ground-based and basic research programs of NSF, and the complementary operational programs of other agencies such as NOAA, DOD, and DOE. During this study, the community will survey solar and space physics and recommend priorities for the decade 2003-2013. Attention will be given to effective implementation of proposed and existing programs and to the human resource aspects of the field involving education, career opportunities, and public outreach. Promising areas for the development of new technologies will be suggested. A minor but important part of the study will be the review of complementary initiatives of other nations in order to identify potential cooperative programs.

An important aspect of the study's consideration of operational programs will be an assessment of how the research programs of NASA and NSF can serve both to provide the operational tools of agencies such as NOAA and DOD and to provide training for future expert staff for those agencies. The study will consider how the science of solar and space physics can lead to new forecast tools and products that have the potential of making the space weather program more operational, and it will identify appropriate next steps to accomplish the transition from research to operations.

Three of the five panels will be organized around interdisciplinary science themes:

- magnetosphere-ionosphere-atmosphere interactions,
- solar wind-magnetosphere interactions, and
- solar and heliospheric physics.

Each of these panels will consider theory and computation as well as ground-based and space-based research. The first two panels will cover both terrestrial and planetary objectives. The three science panels will be complemented by two cross-disciplinary panels:

- theory, computation, and data exploration, and
- education and society.

The survey committee will be responsible for preparing a summary report. The reports of the study panels along with the summary report will be published by the National Research Council. One important goal of these reports is to address the scientific foundation and priorities for the implementation of major NASA programs such as Living with a Star, Solar-Terrestrial Probes, Solar Probe, and Interstellar Probe; and major NSF facilities such as the Relocatable Radar.

In conducting its work, the CSSP would draw on an extensive history of prior studies performed by the Space Studies Board, including:

- *Astronomy and Astrophysics in the New Millennium* (Astronomy and Astrophysics Survey Report [2001]) and *Astronomy and Astrophysics in the New Millennium: Panel Reports* (2002). (Survey and panel reports are joint projects of the SSB and the NRC Board on Astronomy and Astrophysics.)
- *Readiness for the Upcoming Solar Maximum* (1998).
- *Ground-Based Solar Research: An Assessment and Strategy for the Future* (1998).
- *Scientific Assessment of NASA's SMEX and MIDEX Space Physics Mission Selections* (1997).
- *An Assessment of the Solar and Space Physics Aspects of NASA's Space Science Enterprise Strategic Plan* (1997).
- *Space Weather: A Research Briefing* (Web report, 1997).
- *A Science Strategy for Space Physics* (1995).

B

Acronyms and Abbreviations

ACE	Advanced Composition Explorer
ACR	anomalous cosmic ray
AFOSR	Air Force Office of Scientific Research
AIM	atmosphere-ionosphere-magnetosphere
AMISR	Advanced Modular Incoherent Scatter Radar
ATST	Advanced Technology Solar Telescope
AU	astronomical unit (~150,000,000 km)
AURA	Associated Universities for Research in Astronomy
CEDAR	Coupling, Energetics, and Dynamics of Atmospheric Regions
CIR	corotating interaction region
CME	coronal mass ejection
COTS	commercial off-the-shelf
DOD	Department of Defense
EHR	Directorate of Education and Human Resources (NSF)
ESA	European Space Agency
EUV	extreme ultraviolet
eV	electron volt
FASR	Frequency-Agile Solar Radiotelescope
FAST	Fast Auroral Snapshot Explorer
GCR	galactic cosmic ray
GEC	Geospace Electrodynamic Connections
GEM	Geospace Environment Modeling
GEO	Directorate of Geosciences (NSF)

GOES	Geostationary Operational Environmental Satellite
GONG	Global Oscillations Network Group
GS	Geospace
IMAGE	Imager for Magnetopause-to-Aurora Global Exploration
IMF	interplanetary magnetic field
IMP-8	Interplanetary Monitoring Platform (8)
ISAS	Institute of Space and Astronautical Science (Japan)
ISM	interstellar medium; also, Interstellar Sampler Mission
ISP	In-Space Propulsion
ITAR	International Traffic in Arms Regulations
JPM	Jupiter Polar Mission
L1	Lagrangian point 1
lidar	light detection and ranging
LISM	local interstellar medium
LWS	Living With a Star (NASA)
MagCon	Magnetospheric Constellation
MDI	Michelson Doppler Imager
MEMS	microelectromechanical systems
MESSENGER	Mercury Surface, Space Environment, Geochemistry, and Ranging
MHD	magnetohydrodynamic
MHM	Multispacecraft Heliospheric Mission
MIDEX	Medium-Class Explorer
MMS	Magnetospheric Multiscale
MO&DA	mission operations and data analysis
MRX	Magnetic Reconnection Experiment
NASA	National Aeronautics and Space Administration
NOAA	National Oceanic and Atmospheric Administration
NRA	NASA research announcement
NRC	National Research Council
NSF	National Science Foundation
NSO	National Solar Observatory
NSWP	National Space Weather Program

ONR	Office of Naval Research
OSS	Office of Space Science (NASA)
PI	principal investigator
PIDDP	Planetary Instrument Definition and Development Program
PSBL	plasma sheet boundary layer
PVO	Pioneer Venus Orbiter
RAO	Relocatable Atmospheric Observatory
RTG	radioisotope thermoelectric generator
SAMPEX	Solar Anomalous and Magnetospheric Particle Explorer
SDO	Solar Dynamics Observatory
SEC	Sun-Earth Connection program
SEP	solar energetic particle
SHINE	Solar, Heliospheric, and Interplanetary Environment
SMEX	Small Explorer
SMI	Stereo Magnetospheric Imager
SO	Solar Orbiter
SOFIA	Stratospheric Observatory for Infrared Astronomy
SOHO	Solar and Heliospheric Observatory
SR&T	Supporting Research and Technology
SSX	Swarthmore Spheromak Experiment
STEP	Solar-Terrestrial Energy Program
STEREO	Solar Terrestrial Relations Observatory
STP	Solar Terrestrial Probes (NASA)
SWS	Solar Wind Sentinels
SXI	Solar X-ray Imager
TEC	total electron content
TIMED	Thermosphere Ionosphere Mesosphere Energetics and Dynamics
TRACE	Transition Region and Coronal Explorer
TWINS	Two Wide-Angle Imaging Neutral-Atom Spectrometers
UA	Upper Atmosphere base program
UNEX	University-Class Explorer
UV	ultraviolet

C
Biographical Information for Members of the Solar and Space Physics Survey Committee

LOUIS J. LANZEROTTI, *Chair*, is distinguished member of the technical staff, Bell Laboratories, Lucent Technologies. Dr. Lanzerotti's principal research interests have included space plasmas, geophysics, and engineering problems related to the impact of space processes on space and terrestrial technologies. He has been a coinvestigator and principal investigator on several NASA missions, including Galileo and Ulysses, and has conducted extensive ground-based and laboratory research on space and geophysics topics. He was chair (1984-1988) of NASA's Space and Earth Science Advisory Committee and a member of the 1990 Advisory Committee on the Future of the U.S. Space Program. He has also served as chair (1988-1994) of the Space Studies Board of the National Research Council (NRC) and as a member (1991-1993) of the Vice President's Space Policy Advisory Board. He has served on numerous NASA, National Science Foundation, and university advisory bodies concerned with space and geophysics research. He is a member of the International Academy of Astronautics and is a fellow of the Institute of Electrical and Electronics Engineers, the American Geophysical Union, the American Institute of Aeronautics and Astronautics, the American Physical Society, and the American Association for the Advancement of Science. Dr. Lanzerotti is a member of the National Academy of Engineering.

ROGER L. ARNOLDY is a research professor at the University of New Hampshire (UNH). Until his recent retirement, he directed the UNH Space Science Center and was a professor in the UNH Department of Physics. Dr. Arnoldy's research interests are in space physics, the space plasma physics of the solar wind's interaction with Earth's magnetic field, and auroral particle acceleration. He has served as a principal investigator for the Arctic and Antarctic magnetic pulsation studies and for more than 30 auroral sounding rocket flights. Dr Arnoldy was an investigator on some of the earliest U.S. space missions, including Explorer VI launched in 1959—the first satellite to be orbited by NASA. He has been coinvestigator for the ECHO rocket flights and the Antarctic Penguin Automated Geophysical Observatories. Dr. Arnoldy has been a trustee of the University Space Research Association and a member of the NASA Sounding Rocket Users Committee and of the Peer Review Panel of the NASA Wallops Flight Center. Dr. Arnoldy is a fellow of the American Geophysical Union. He has more than 100 publications in refereed space physics journals.

FRAN BAGENAL is a professor of astrophysical and planetary sciences and a research associate in the Laboratory for Atmospheric and Space Physics at the University of Colorado, Boulder. Her research interests include the synthesis of data analysis and theory in the study of space plasmas. She specializes in the field of planetary magnetospheres, particularly jovian magnetospheres, and solar corona. Dr. Bagenal has received six NASA Group Achievement Awards in the past 20 years. She is currently an interdisciplinary scientist for the Galileo project and a member of the New Horizons science team. She is a member of the American Astronomical Society, the American Geophysical Union, the American Physical Society, the Royal Astronomical Society, and the American Association of Physics Teachers. Dr. Bagenal was a member of the NRC Space Studies Board from 1998 to 2001.

DANIEL N. BAKER is the director of the Laboratory for Atmospheric and Space Physics, University of Colorado, and is a professor of astrophysical and planetary sciences. His primary research interest is the study of plasma physical and energetic particle phenomena in the planetary magnetospheres and in Earth's magnetosphere, and he conducts research in space instrument design, space physics data analysis, and magnetospheric modeling. He joined the physics research staff at Los Alamos National Laboratory and became leader of the Space Plasma Physics Group in 1981. From 1987 to 1994, he was the chief of the Laboratory for Extraterrestrial Physics at NASA Goddard Space Flight Center. Dr. Baker has published more than 500 papers in the refereed literature and is a fellow of the American Geophysical Union and the International Academy of Astronautics. He currently serves on the NRC U.S. National Committee for the International Union of Geodesy and Geophysics. He has served on numerous NRC committees and panels, including the Committee on Data Management and Computation, 1986-1987, the Committee on Solar and Space Physics, 1984-1986, and the Space Studies Board, 1995-2000.

JAMES L. BURCH is vice president, Southwest Research Institute (SwRI), Space Science and Engineering Division. Dr. Burch was a space physicist at NASA for 6 years prior to his going to SwRI in 1977. In 1996, Dr. Burch was selected as the principal investigator for the NASA Imager for Magnetopause-to-Aurora Global Exploration (IMAGE) mission, which is providing the first-ever global images of key regions of Earth's magnetosphere as they respond to variations in the solar wind. Dr. Burch was elected a fellow of the AGU in recognition of his work in the field of space physics and aeronomy, including research on the interaction of the solar wind with Earth's magnetosphere and the physics of the aurora. Dr. Burch has served as editor in chief of *Geophysical Research Letters* and as president-elect and president of the Space Physics and Aeronomy Section of the AGU. He currently serves on the governing board of the American Institute of Physics (AIP) and chairs the NRC Committee on Solar and Space Physics.

ARTHUR CHARO, *Study Director*, received his Ph.D. in physics from Duke University in 1981 and was a postdoctoral fellow in chemical physics at Harvard University from 1982 to 1985. Dr. Charo then pursued his interests in national security and arms control at Harvard University's Center for Science and International Affairs, where he was a fellow from 1985 to 1988. From 1988 to 1995, he worked in the International Security and Space Program in the U.S. Congress's Office of Technology Assessment (OTA). He has been a senior program officer at the Space Studies Board (SSB) of the National Research Council since OTA's closure in 1995. Dr. Charo is a recipient of a MacArthur Foundation Fellowship in International Security (1985-1987) and was the American Institute of Physics Congressional Science Fellow for 1988-1989. He is the author of research papers in the field of molecular spectroscopy; reports on arms control, Earth remote sensing, and space policy; and a monograph, *Continental Air Defense: A Neglected Dimension of Strategic Defense* (University Press of America, 1990).

JOHN C. FOSTER is associate director of the Haystack Observatory at the Massachusetts Institute of Technology (MIT); head of the Atmospheric Sciences Group, Millstone Hill Observatory; and MIT principal research scientist. His research interests are in the physics of the magnetosphere, ionosphere, and thermosphere. Topics of particular interest to him include magnetosphere-ionosphere-atmosphere coupling; incoherent scatter radar; plasma waves and instabilities; ionospheric convection electric fields; and midlatitude/inner-magnetosphere phenomena. Dr. Foster has also engaged in scientific collaboration with investigators at international research facilities. He is currently a member of the NRC Committee on Solar and Space Physics and is a past member of the NRC Board on Atmospheric Sciences and Climate's Committee on Solar-Terrestrial Research, and of the NRC U.S. National Committee for the International Union of Radio Science.

PHILIP R. GOODE is director of the Big Bear Solar Observatory (BBSO) at Big Bear Lake, California; director of the Center for Solar-Terrestrial Research at the New Jersey Institute of Technology (NJIT); distinguished professor of physics and mathematics at the NJIT; and visiting associate in physics, mathematics, and astronomy at the California Institute of Technology. Dr. Goode has held research positions at Rutgers University, the University of Rochester, and the University of Arizona and was a member of the technical staff at Bell Labs. His primary research interests are the internal structure of the Sun; the nature of the Sun's magnetic fields, flares, and coronal mass ejections; and space weather. He is also measuring and modeling Earth's reflectance from studies of earthshine and satellite cloud cover data. Dr. Goode was a member of the NRC Panels on Management and Research in Astronomy and Astrophysics (2000, "Augustine/Blue Ribbon Panel") and on Solar Astronomy (1998-2000), which advised the Astronomy and Astrophysics Survey Committee on scientific opportunities and priorities in the field of solar astronomy.

RODERICK A. HEELIS is the Cecil and Ida Green Honors Professor of Physics and director of the William B. Hanson Center for Space Sciences at the University of Texas at Dallas. His research specialization covers planetary atmospheres, ionospheres, and magnetospheres and the physical phenomena coupling these regions. He serves as a principal investigator for grant and contract research sponsored by DOD, NASA, and NSF. Dr. Heelis is currently a member of the NASA Space Science Advisory Committee and has served on the NRC Committee on Solar and Space Physics.

MARGARET G. KIVELSON is a professor of space physics in the Department of Earth and Space Sciences and the Institute of Geophysics and Planetary Physics (Space Science Center) at the University of California, Los Angeles. Her principal scientific interests are magnetospheric plasma physics of Earth and Jupiter (theory and data analysis) and interaction of flowing plasmas with planets and moons (theory, data collection, and data analysis). She is currently active in space projects including Galileo (magnetometer principal investigator), Polar (coinvestigator), and Cluster (coinvestigator). Dr. Kivelson is currently a member of the NRC Space Studies Board, and she is a former member of the NRC Commission on Physical Sciences, Mathematics, and Applications and of the Committee on Solar and Space Physics. She is a member of the American Academy of Arts and Sciences and is a fellow of the American Institute of Aeronautics and Astronautics, the American Geophysical Union, the American Physical Society, and the American Association for the Advancement of Science. Dr. Kivelson is a member of the National Academy of Sciences.

WILLIAM S. LEWIS, *Consultant*, received his Ph.D. in German literature from Rice University in 1987. Since 1988, he has worked in the Space Science Department at Southwest Research Institute (SwRI) in San Antonio, Texas. Dr. Lewis is currently principal scientist in SwRI's Space Science and Engineering Division. His primary research interests are the jovian and terrestrial auroras. He is currently working on the analysis of data obtained with the far-ultraviolet imager on the IMAGE spacecraft, with particular emphasis on the proton aurora.

WILLIAM H. MATTHAEUS is a professor in the Bartol Research Institute at the University of Delaware. His major areas of interest include the characterization of interplanetary plasma turbulence, the identification of nonlinear dynamical processes in the solar wind using data from spacecraft observations, and the study of coronal heating mechanisms. Theoretical and computational studies of turbulent magnetohydrodynamic plasmas have also continued to be an emphasis of his work. Dr. Matthaeus is a member of NASA's Sun-Earth Connections Advisory Subcommittee (SECAS). He is a fellow of the American Geophysical Union and the American Physical Society and is a recipient of the AGU's MacElwane Prize.

FRANK B. McDONALD is senior research scientist, Institute for Physical Science and Technology, University of Maryland. His current research is directed toward understanding the properties of the galactic cosmic radiation, its transport in the interstellar medium, and its modulation by our own heliosphere; the acceleration and transport of solar/interplanetary energetic particles; and the study of the dynamics of the outer heliosphere. Dr. McDonald's research makes extensive use of the energetic particle data from Pioneer 10 and 11, on which he is the principal investigator for one of the cosmic-ray experiments, and on the Voyager cosmic-ray experiment, for which he is a coinvestigator. Dr. McDonald was chief of Goddard's Laboratory for High Energy Astrophysics, NASA chief scientist, and associate director of the Goddard Space Flight Center. He has served on several NRC committees related to space research and is currently a member of the NRC Committee on Solar and Space Physics. Dr. McDonald is a member of the National Academy of Sciences.

EUGENE N. PARKER is the S. Chandrasekhar Distinguished Service Professor Emeritus, Departments of Astronomy and Astrophysics, and Physics, University of Chicago. Dr. Parker is one of the nation's most distinguished theoretical astrophysicists and is a recipient of numerous prizes from his peers. His extensive NRC service includes serving as chair of the NAS Astronomy Section from 1983 to 1986. Dr. Parker's current research interests include theoretical plasma physics; magnetohydrodynamics; solar and terrestrial physics; basic physics of the active star; application and extension of classical physics to the active conditions found in the astronomical universe (e.g., the stellar x-ray corona); and the solar wind and the origin of stellar and galactic magnetic fields. Dr. Parker served as chair of the NRC Task Group on Ground-Based Solar Research (1997-1998) and is currently a member of the NRC Committee on Solar and Space Physics. Dr. Parker is a member of the National Academy of Sciences.

GEORGE C. REID currently holds an appointment as a senior research scientist in the University of Colorado's Cooperative Institute for Research in Environmental Sciences in Boulder. His research activities have included the physics of solar energetic particles and their interaction with the atmosphere, the dynamics and chemistry of the middle atmosphere, and most recently the mechanisms of global climate change, with emphasis on the tropics and including especially the influence of solar variability. Dr. Reid's career includes various positions with the NOAA Aeronomy Laboratory (1970-1998) and Space Environment Laboratory (1963-1970) in Boulder, Colorado. He has also served as Associate Professor of Geophysical Research at the University of Alaska (1958-1960) and in adjunct positions at the University of Colorado and Colorado State University. He has served on numerous committees of the NRC, as editor-in-chief of the *Journal of Geophysical Research* (Space Physics), and as president of the Space Physics and Aeronomy section of the American Geophysical Union. He is a fellow of AGU and a recipient of the U.S. Commerce Department's Gold Medal.

ROBERT W. SCHUNK is a professor and the director of the Center for Atmospheric and Space Science, Utah State University. His expertise is in plasma physics, fluid mechanics, aeronomy, space physics, electricity and magnetism, and data analysis. Dr. Schunk has developed numerous computer models of space physics phenomena, regions, and spacecraft-environment interactions. He has published more than 300 papers in the refereed literature, comparing model predictions with measurements, with many of them using data from several coherent and incoherent scatter radars, ionosondes, rockets, satellites, and the space shuttle. He is vice chair of Commission C of COSPAR, a vice chair of Division II of IAGA, and a member of Commissions G and H of the International Union of Radio Science. Dr. Schunk has served on several NRC committees related to space science and is currently a member of the Committee on Solar and Space Physics. He received the Governor's Medal for Science and Technology from the State of Utah and is a fellow of the American Geophysical Union.

ALAN M. TITLE is a senior fellow at the Lockheed Martin Space Systems Advanced Technology Center in Palo Alto, a consulting professor of physics at Stanford University, and co-director of the Stanford-Lockheed Institute for Space Research. Dr. Title's primary research areas are solar magnetic and velocity fields; optical interferometers, in particular ultranarrow optical filters; high-resolution observations using active and adaptive optical systems; and data analysis systems for image analysis. Dr. Title is the principal investigator for the Transition Region and Coronal Explorer (TRACE) mission, which is making space-based observations of the Sun to study the connection between its magnetic fields and the heating of the corona. Dr. Title was a member of the most recent NRC decadal Astronomy and Astrophysics Survey Committee and served on the NRC Panel for Review of the Explorer Program. He also served as a member of the NRC Committee on Solar and Space Physics and of the Space Studies Board. Dr. Title is the recipient in 2001 of the George Ellery Hale Prize of the Solar Physics Division of the American Astronomical Society. Dr. Title is a member of the National Academy of Engineering.